Enterobacteria

Edited by Sonia Bhonchal Bhardwaj

Published in London, United Kingdom

Enterobacteria
http://dx.doi.org/10.5772/intechopen.98173
Edited by Sonia Bhonchal Bhardwaj

Contributors
Navneet Kaur, Ashwini Agarwal, Malika Grover, Sanampreet Singh, Bianca Mendes Maciel, Eliege Jullia Eudoxia dos Santos, Amanda Teixeira Sampaio Lopes, Mashooq Ahmad Dar, Kaisar Ahmad Bhat, Tasaduq Manzoor, Asmat Farooq, Kaisar Ahmad Allie, Shaheen Majeed Wani, Tashook Ahmad Dar, Ali Asghar Shah, Daoudou Bakari, Jojo Mbala Ekamba, Mamoudou Hamadou, Yaya Garga, Martin-Paul Baane, Félicite Obono, Mohd Afendy Abdul Talib, Son Radu, Cheah Yoke Kqueen, Farinazleen Mohamad Ghazali, Ping Li, Han Jiang, Jiayi Xiong, Mengqi Fu, Xianpu Huang, Boxun Huang, Qing Gu, Sonia Bhonchal Bhardwaj

First published in London, United Kingdom, 2022 by IntechOpen
IntechOpen is the global imprint of INTECHOPEN LIMITED, registered in England and Wales, registration number: 11086078, 5 Princes Gate Court, London, SW7 2QJ, United Kingdom

British Library Cataloguing-in-Publication Data
A catalogue record for this book is available from the British Library

Additional hard and PDF copies can be obtained from orders@intechopen.com

Enterobacteria
Edited by Sonia Bhonchal Bhardwaj
p. cm.
Print ISBN 978-1-80355-309-2
Online ISBN 978-1-80355-310-8
eBook (PDF) ISBN 978-1-80355-311-5

Meet the editor

Dr. Sonia Bhonchal Bhardwaj, Ph.D., is a senior assistant professor at the Department of Microbiology, Dr. Harvansh Singh Judge Institute of Dental Sciences and Hospital, Panjab University, Chandigarh, India. She has published several international publications in reputed journals as well as books, book chapters, and congress proceedings. Dr. Bhardwaj has received grants from the Department of Science and Technology, India, and has worked on biofilm formation in *Enterococcus faecalis* in periodontitis, *Streptococcus mutans*, and phage therapy. She is a member of the Indian Association of Medical Microbiology (IAMM), Gastrointestinal Infection Society of India (GISI), Association of Microbiologists of India (AMI), and Society for Bacteriophage Therapy (SBRT).

Contents

Preface

Enterobacteriaceae are a large family of Gram-negative, non-sporing bacilli. These bacteria are a part of the natural flora of animals including humans, but some pathogenic species are associated with intestinal and extra-intestinal diseases. They are a common cause of urinary tract infections (UTIs) and diarrhea and can enter the bloodstream causing life-threatening complications. Antibiotic resistance is also expressed as a common characteristic among enterobacteria globally, as UTIs and recurrent UTIs are associated with significant use of antibiotics promoting multi-drug resistance. These bacteria can synthesize several enzymes such as extended-spectrum β-lactamases, carbapenemases, metallo-β- lactamases, and many others. It is hard to treat the strains that are resistant to antibiotics due to the cause of recurrent and untreatable infections.

This book describes important enterobacteria including *Escherichia coli, Klebsiella, Salmonella, Proteus, Shigella, Serratia, Enterobacter, Citrobacter*, and other species. Chapters address the identification, classification, and pathogenicity of enterobacteria as well as the role of pathogenic enterobacteria in infection including UTIs and food poisoning. Its presence in wild animals is also discussed. To overcome the problem of antibiotic resistance, alternative treatment strategies like bacteriophages are gaining popularity. As such, this book also examines the role of bacteriophages in the management of UTIs caused by enterobacteria.

Written by experts from all over the world, this book provides a comprehensive overview of enterobacteria and their pathogenesis, virulence factors, and treatment strategies as well as their link to multi-drug resistance. It is a useful resource for medical microbiologists, clinicians, researchers, and students interested in the study of enterobacteria.

I would like to wholeheartedly thank all the authors for their excellent contributions. I am also grateful to Author Service Manager Sara Debeuc and the staff at IntechOpen for their concern and encouragement in publishing this book.

Dr. Sonia Bhonchal Bhardwaj
Senior Assistant Professor,
Department of Microbiology,
Dr. Harvansh Singh Judge Institute of Dental Sciences and Hospital,
Panjab University,
Chandigarh, India

Section 1

Escherichia Coli

Chapter 1

Uropathogenic *Escherichia coli*

Navneet Kaur, Ashwini Agarwal, Malika Grover and Sanampreet Singh

Abstract

Urinary tract infections (UTIs) are among the most common infections encountered worldwide in clinical practice. *Escherichia coli* is by far the most frequent cause of infections responsible for nearly 80–90% of the infections. The strains of *E. coli* causing UTI are termed as uropathogenic *E. coli*. They vary from commensal strains as they have acquired virulence and resistant determinants through plasmids, bacteriophages, pathogenicity islands or DNA horizontal transfer of transposons which permits them to victoriously colonize the urinary tract and cause a broader spectrum of disease. For the fact, UPEC strains possess an abundance of both structural (as fimbriae, pili, flagella, capsule, lipopolysaccharide) and secreted (toxins, iron-acquisition systems, enzymes) virulence factors that play a crucial role in the pathogenesis. The pathogenesis of UPEC involves adherence, colonization, evading host defenses and damage to host tissue to achieve virulence. UTI is often treated empirically by broad-spectrum antibiotics in the absence of culture and susceptibility results. This over-use of antibiotics has resulted in the development of antibiotic resistance worldwide. Having a detailed understanding of the bacterium and its virulence factors can help us in developing new treatment options in presence of global antimicrobial resistance.

Keywords: UTI, UPEC, virulence factors, adhesins, toxins, antimicrobial resistance

1. Introduction

Urinary tract infections (UTIs) are among the most common infections encountered worldwide in clinical practice accounting for approximately 150 million cases annually causing heavy burden on health infrastructure [1]. Women, undoubtedly are at greatest risk as compared to males and it is been observed that almost 50% of all women have experienced UTI at least once in their lifetime. The infection may result either due to the pathogenicity of the offending microorganism, host susceptibility or a combination of both. While many different microorganisms are known to cause UTI which includes bacteria, viruses and fungi, bacteria remain the main cause responsible for over 95% of cases. Among bacteria, *Escherichia coli (E. coli)* is by far the most frequent cause responsible for nearly 80–90% of the infections [2]. The most common route of infection of *E. coli* is the bacterial colonization of the urethra followed by the ascension to the bladder. Normally, *E. coli* is present as a commensal flora in the lower gastrointestinal tract of humans, nonetheless, there are a few highly adapted *E. coli* clones present that have acquired specific virulence attributes, which gives them an escalated ability to adapt to new niches and permits them to cause a broad spectrum of disease. Uropathogenic *Escherichia coli* (UPEC) is simply the pathotype of extraintestinal pathogenic *E. coli* which were first isolated from the urine of the patients having UTI. They differed from those cultured from the stool specimens of healthy individuals and those causing diarrhea, hence the term UPEC. UPEC strains possess an abundance of both structural (as fimbriae, pili, flagella) and secreted (toxins, iron-acquisition systems) virulence factors that play an important part in the pathogenesis, however its capability to adhere to host epithelial cells in the urinary tract serves as the most important determinant of pathogenicity [3].

2. What is uropathogenic *E. coli*? How it causes infection?

2.1 Uropathogenic *E. coli*

Escherichia coli, a Gram-negative bacilli belonging to the family Enterobacteriaceae resides in the gastrointestinal tract of humans as a part of their microbiota. It normally remains in harmony with its host and seldom causes disease except in an immunocompromised host. However, few strains of *E. coli* can split from their commensal cohort taking on a pathogenic form. That means these strains acquire specific virulence factors through plasmids, pathogenicity islands or DNA horizontal transfer of transposons that bestow them the ability to adjust to new niches and cause a broad spectrum of diseases [4]. Further, only the most successful combinations of virulence factors persist to become specific pathotypes of *E. coli* [4]. The strains of *E. coli* which are pathogenic are divided into diarrheagenic or enteric *E. coli* and extraintestinal *E. coli* (EXPEC) based on the body site they colonize. Diarrheagenic *E. coli* includes enterotoxigenic (ETEC), enteropathogenic (EPEC), shiga toxin-producing (STEC), enteroaggregative (EAEC), enteroinvasive (EIEC), and diffusely adherent *E. coli* (DAEC). While EXPEC is mainly uropathogenic *E. coli* (UPEC), neonatal meningitis *E. coli* (NMEC) and sepsis-associated *E. coli*. Based on phylogenetic analysis using multilocus enzyme electrophoresis, pathogenic *E. coli* pathotypes (from both intestinal and extra-intestinal *E. coli*) were divided into four phylogenetic groups A, B1, B2 and D. These phylogenetic groups are representative of their genetic origin [5].

Uropathogenic *E. coli* as described belongs to the extraintestinal pathogenic *E. coli* group and is associated with a subset of serogroups and serotypes (O1:H4, O1:H6, O1:H7, O1:H$^-$, O2:H1, O2:H4, O4:H5, O6:H1, O7:H4, O7:H6, O7:H$^-$, O18ac:H7, O18ac:H$^-$, O22:H1, O25:H1, O75:H5 & O75:H7) and with the B2 or D phylogenetic groups [6]. UPEC possesses acquired virulence and resistant determinants which permits it to victoriously colonize the urinary tract and cause disease.

2.1.1 CFT073

CFT073 is a prototypical strain of UPEC that was recovered from a woman having severe pyelonephritis infection, it belongs to phylogenetic group B2. It was noted to have increased hemolytic activity in comparison to other UPEC strains. On sequencing, its virulence genes were found to be grouped into five pathogenicity islands [7, 8].

2.2 Route of infection

For UPEC to cause infection, the ultimate origin of it is the intestinal tract of the human host which finally acts as a fecal reservoir. Principally apparent in women, the first step is the bacterial colonization of the vaginal introitus and periurethral meatus. Notably, colonization takes place in parallel with the loss of protective vaginal *Lactobacillus* species. This follows ascension into the bladder and adherence to bladder epithelium (uroepithelium). This is followed by UPEC internalization by umbrella cells (it is the outermost layer of the uroepithelium). Inside the bladder cells, most of the bacteria are exocytosed while the minority of them will evade this mechanism gain entrance into the cytosol to form IBCs (intracellular bacterial communities). When these intracellular bacteria stop replicating, they enter another stage known as QIR (quiescent intracellular reservoir) and are behind recurrent UTI episodes.

In this manner, infection of the lower urinary tract has the power to advance to kidneys and enter the bloodstream to cause urosepsis (**Figures 1** and **2**) [3].

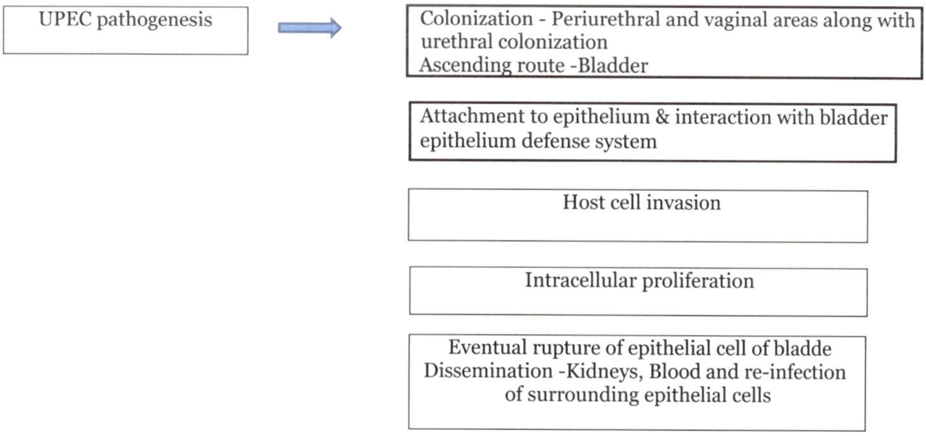

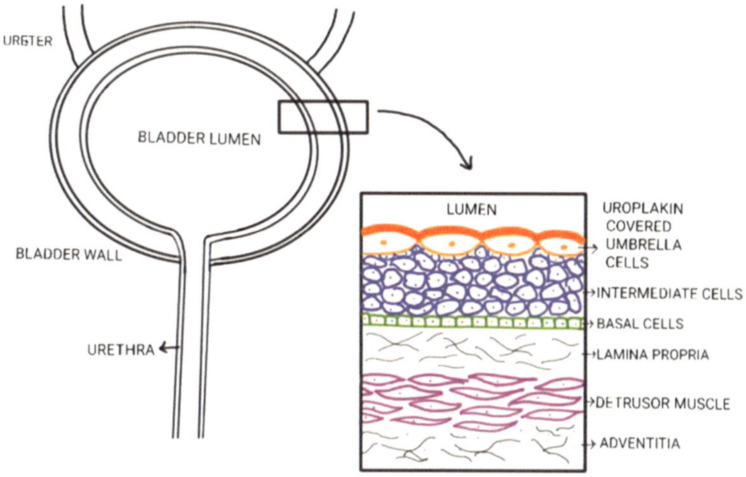

Figure 1.
Structure of the urinary bladder and urothelium. The urothelium (transitional epithelium) is believed to form vital and essential hurdle to infection that includes mucus glycosaminoglycans retarding adherence of UPEC, infection-resistant umbrella cells and glycoprotein plates known as uroplakins.

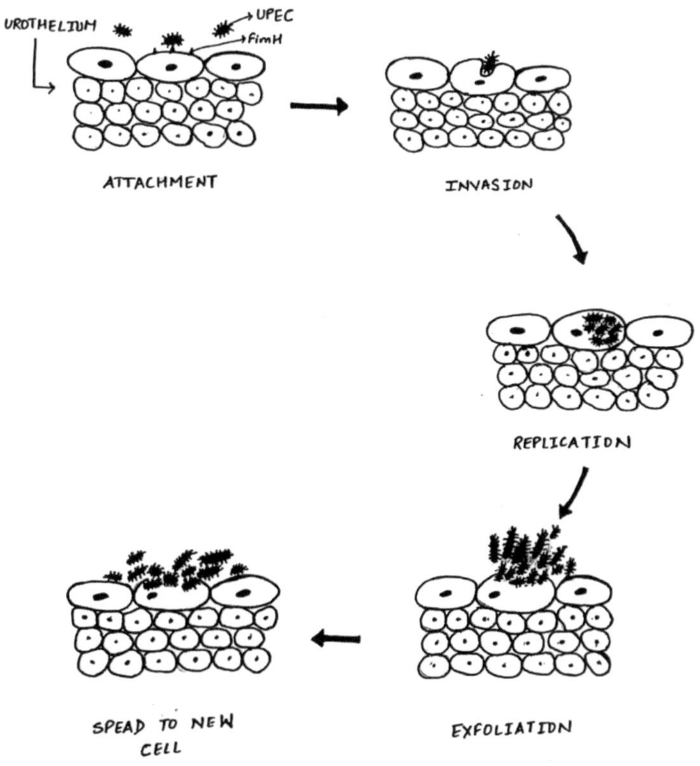

Figure 2.
UPEC pathogenesis. UPEC expresses pili systems (Fim H) for adherence to the epithelial cells of the bladder. It follows invasion into the host cell which initiates replication to form IBCs and a subpopulation also undergoes cell elongation (filamentation). Ultimately the epithelial cell is overloaded and UPEC escapes, rupturing open the host cell releasing motile short and elongated cells which can infect neighboring host epithelia to continue the infective cycle.

3. Bladder defenses against UPEC

To cause infection, UPEC has to beat the natural urinary tract defenses that include physical and chemical defenses. Physical defenses include high urine flow, exfoliation of cells, high urine osmolality and low pH. Chemical defenses of the

Bacterial aim	Host barrier
Attachment to the host cell surface with the help of adhesins (P fimbriae, type 1 fimbriae)	Flow of urine, mucociliary blanket
Acquisition of nutrients by cellular lysis by hemolysin, iron acquisition by siderophores	Sequestration of nutrients (iron through intracellular storage)
Initial avoidance of bactericidal activity by the host—capsular polysaccharides and lipopolysaccharide	Phagocytic cells, complement, antimicrobial peptides
Late avoidance of bactericidal activity by the host—Antimicrobial resistance	Antimicrobial therapy, acquired immunity

Table 1.
Interaction of UPEC with the human host.

urinary tract comprise secreted proteins such as THP (Tamm-Horsfall protein), IgA (immunoglobulin), antimicrobial peptides and immune system activation (innate and adaptive). **Table 1** summarizes the interaction of UPEC with the human host.

4. Virulence factors of UPEC

There are various virulence factors possessed by UPEC which enable its colonization and pathogenicity. These factors are inserted either in the genome (chromosome) by insertion elements or transposons or encoded on plasmids. The virulence factors are associated with the bacterial cell surface such as adhesins and factors that are secreted and carried to the site of action such as toxins [9].

4.1 Adhesins

For UPEC to colonize and cause infection, it needs to be equipped with some adhesive molecules which will promote attachment to the host cell surface. Adherence to host cell is regarded as the crucial step for colonization because in normal instances regular urinary flow does not permit colonization of the bacteria to the urinary tract. The adhesive factors found in UPEC are known to be adhesins which exists either in the form of filamentous surface organelles called pili or fimbriae or as non-filamentous proteins in the outer membrane.

4.1.1 P fimbriae

Edén and his colleagues discovered in the year 1976 when they identified that *E. coli* that was cultured from pyelonephritis cases attached in more numbers to exfoliated uroepithelial cells when compared to *E. coli* strains obtained from fecal samples [10]. This strong adherence was linked to the presence of 'fimbriae' which when isolated could also specifically adhere to the uroepithelial cell surface [11]. Those bacteria which were expressing this type of fimbria could agglutinate human O erythrocytes and the hemagglutination wasn't been able to be inhibited by mannose hence Mannose Resistant. This kind of agglutination made these new fimbriae distinct from type 1 fimbriae. Later work revealed the receptor to which these new fimbriae i.e., 'P fimbriae' binds is globoseries receptor which is a component of P blood group antigen found in human erythrocytes and uroepithelial cells. Furthermore, this antigen was identified to be a glycospingolipid (synthesized by specific glycosyltransferases and constitutent of glycocalyx surrounding the uroepithelial cells) with a lipid moiety anchored in the cell membrane and a chain of carbohydrates exposed on the erythrocyte surface. These globoseries glycosphingolipid receptors (Gal-Gal) are spread evenly all over the urinary tract especially in kidneys.

P fimbriae are said to increase UPEC virulence at various stages of infection. They help the bacteria to persist longer in the intestinal tract and expand more strongly in the urinary tract with the plan of colonization and going ahead with ascending infection [12, 13]. So, when they reach the urinary tract, *E. coli* strains having P fimbriae attach, persists and even in the presence of enhanced immune response (engaging toll-like receptors 4 and cytokine elaboration) invades kidneys and can cause bacteremia. For the reason, there is a correlation between P fimbriae and acute disease severity in more than 90% of cases. However, <20% asymptomatic carriers also express this P fimbriae. The adhesin complex is encoded by *pap* gene *EFG* sequences.

There exist 3 molecular variants of PapG adhesin encoded by *Pap*G class I through IV alleles, these have different receptor binding preferences ultimately affecting clinical outcomes. For example, allele class II is predominant among strains causing pyelonephritis and bacteremia whereas class III is frequently encountered in women having cystitis and children [14].

4.1.2 Type I fimbriae

Type I fimbriae are considered as a crucial virulent factor in UTI but their exact individual role is challenging to understand as they are expressed by pathogenic as well as commensal strains of *E. coli* including other genera within family enterobacteriaceae. Additionally, it is found that there is an absence of any significant difference in the frequency of the *fim* gene (which encodes type I fimbriae) among more or less virulent strains [15]. These fimbriae are encoded by an operon that contains nine genes present on the chromosome of most of the UPEC in the order of: *fimB, fimE, fimA, fimI, fimC, fimD, fimF, fimG, fimH;* these encoding structural and regulatory proteins [3].

Type I fimbriae bind to uroplakin Ia and IIIa (urothelial mannosylated glycoprotein) through the *FimH* subunit [4]. *FimH* is a tip protein of type I fimbriae. In addition, it may also bind to other cell-surface proteins such as integrins, fibronectin, Tamm-Horsfall protein (THP), etc. This binding or interaction results in molecular phosphorylation events that are necessary for the stimulation of signaling pathways involved in invasion and apoptosis.

Besides, playing a role in attachment to bladder epithelium, it is also found to directly trigger invasion by UPEC into epithelial cells of the bladder (BECs) where they induce formation of IBCs and remain as reservoirs to act as a source of clinical relapse [16, 17]. Some studies have given an insight to the fact that type I fimbriae enhance the infectious potential of UPEC [18, 19] but the accurate timing of expression of fimbria during urinary tract infection remains blurred. It was also seen that UPEC isolates obtained from clinical samples (urine) during infection expressed little to no type I fimbriae [20]. In addition, fimbrial expression was more or less absent in UPEC strains from the urine of women with cystitis [21].

The expression of these fimbriae is finely regulated attending to environmental signals and is under the control of phase variation that determines the percentage of fimbriated cells in the population.

Hultgren and colleagues, in the experimental murine model of UTI, found out that *E. coli* which was obtained from bladder lumen did not express type I fimbriae, however, bacteria that were adhered to the bladder wall did express them. Therefore, their contribution in adherence and colonization cannot be effectively determined by measuring the expression of fimbriae in the bladder lumen [22]. There is also an observation that type I fimbriae are not especially prevalent in pyelonephritogenic strains and adherence of bacteria to urinary catheters is also type I fimbriae dependent.

4.1.3 Dr adhesins

The Dr adhesin family consists of both fimbrial and afimbrial adhesins on *E. coli* surface. There are four genes i.e., *dra A, B, C, D* which encode for adhesins and structural proteins. These adhesins can bind to Dr blood group antigen (a component of decay-accelerating factor which prevents lysis by complement). Inside the urinary tract, they attach to the epithelium of the bladder and type IV collagen present on the basement membranes. Although, these adhesins are present in less number in UTI-causing strains, however collective data from various sources shows that the genes encoding for Dr adhesin family are widespread among cystitis and pyelonephritis strains when compared to control strains (fecal isolates). In the experiments involving mouse models, Dr adesins exhibits tropism for the basement membrane of the renal intersititum, hence integral for chronic pyelonephritis development. Their presence has been linked to epithelial invasion. Some studies in rat models also points out that their interaction with the host cell receptors in kidneys is very much persistent. In addition, it also has a role in the pathogenesis of UPEC as evident in a mouse model study where Dr-positive strain leads to a disease pathologically similar to chronic tubulointerstitial nephritis and a Dr-negative isogenic mutant causes no disease. Type I fimbriae and Dr adhesin together are associated with invasion of epithelial cells of bladder along with intracellular persistence by UPEC [23, 24].

4.1.4 S and F1C fimbriae

S and F1C fimbriae are also involved in the urinary tract infection process. Both fimbriae are shown to have very related biogenesis genes however they have different adhesin alleles. They exhibit binding to epithelial and endothelial cells from the lower urinary tract and kidneys in humans [25, 26]. S fimbriae binds to sialic acid epitopes which are present in renal sialylated lipoproteins. Also, they particularly are also responsible for other extraintestinal infections such as sepsis, meningitis apart from UTI because they may promote the dissemination of bacteria within the host tissues. As per various pooled studies, F1C fimbriae are more usually seen in strains from pyelonephritis and cystitis patients than fecal strains used as controls [27]. Furthermore, experiments conducted in a try to know their exact role in UTI have not been reported clearly but in strain CFT073, in which type 1 and P fimbriae encoding genes have been inactivated, F1C fimbriae were expressed at elevated levels demonstrating a synchronize fimbrial expression in UPEC [28].

4.1.5 Other adhesins

Adhesins belonging to the Afa family are also involved in urinary tract infections. The UPEC strains expressing them have distinctive renal tissue tropism. Further findings suggest that these adhesins have properties that supports the development of chronic or/and recurrent infections [4]. Other adhesins identified in vivo among UPEC strains are type 1c, G, M, X adhesins which vary in molecular binding specificities and serologic properties. **Table 2** summarizes various UPEC adhesins contributing to virulence.

4.2 Toxins

There are three main types of toxins that are secreted by UPEC. These are:

A. Hemolysin

B. Cytotoxic necrotizing factor 1 (CNF)

C. Autotransporters

Type of adhesin	Corresponding receptor	Encoding genes	Special points
Type 1 fimbriae	PMNs and epithelial cells with mannose proteins	*fim* B, *fim* E, *fim* H and *Pil*	—
Type 1c fimbriae	Unknown	*Foc*	—
Type 3 fimbriae	M blood group	mrkABCDF	Mediate formation of biofilm
P fimbriae	P blood group antigen: Gal-α 1–4	*PapG*, *pap*GAP	Associated with bacteremia, cystitis and pyelonephritis
S/F1C fimbriae	Sialyl-α-2-3 galactoside	*Sfa*/*fac*	THP inhibits the adherence
G fimbriae	Terminal N-acety-D-glucosamine	—	—
M fimbriae	Galactose-N-acetylgalactosamine	—	—
Dr family	Type 4 collagen & Dr blood group antigen	*AfaE1–5*, *AfaF*, *Drb* operon	—

Table 2.
Summary of various adhesins of UPEC.

Apart from these, there are other toxins identified which are secreted by UPEC strains and have cytotoxic activity. Toxins production by UPEC leads to inflammatory response causing UTI symptoms.

4.2.1 Hemolysin

In 1921, Dudgeon *et al.*, documented that 50% of *E. coli* isolates which caused UTI were causing hemolysis on blood agar plates in comparison with 13% of fecal isolates. This action was credited to the hemolysin protein secreted by *E. coli* (belonging to the family of RTX toxins (repeats-in-toxin)) [29]. α-hemolysin (Hly A) is the most vital virulence factor secreted by UPEC that is associated with pyelonephritis. The *hly* genes encode for proteins that are needed to synthesize and secrete hemolysin. This toxin shows dual activity dependent on concentration i.e., low and high concentration:

- At low concentrations, it can cause apoptosis of target host cells which involves neutrophils, T-lymphocytes and renal cells and also stimulate the exfoliation of epithelial cells of the bladder [30].

- At high concentrations, HlyA can lyse erythrocytes and nucleated host cells including uroepithelial cells. This may enable UPEC to cross mucosal barriers effectively, destroy effector immune cells and gain advanced access to nutrients and iron stores of the host.

α-Hemolysin can also result in the elevated elaboration of IL-6 and IL-8 by inducing Ca^{2+} oscillations in renal epithelial cells. In addition, this toxin is associated with renal complications in 50% of cases of pyelonephritis and also causes endothelial damage and renal vasoconstriction. To add up permanent renal scarring is a usual complication that follows infection by HlyA *E. coli* [31].

4.2.2 CNF-1

CNF-1 is frequently detected in *E. coli* strains causing UTI and almost always in association with hemolysin with which it is linked genetically. This protein is secreted by

E. coli in vitro and prompts actin stress fibers formation and membrane ruffle formation in a Rho GTPase dependant manner. Various studies describe its potential role in UPEC pathogenesis. CNF-1 appears to increase the attachment of PMNs (polymorphonuclear leukocytes) to T84 monolayers (epithelial cells) theraby decreasing their phagocytic effect. Besides, it leads to apoptosis in the 5637-bladder cell line, an event that might be elucidate the exfoliation of bladder epithelial cells after UPEC infection [32].

4.2.3 Autotransporters

Autotransporters toxins also named as type V secretion toxins consists of SAT (secreted autotransporter toxin) and VAT (vacuolating autotransporter toxin) encoded by UPEC [33]. SAT is demonstrated more commonly among *E. coli* strains (55% strains) associated with pyelonephritis relative to fecal strains (22%). SAT which was first isolated from *E. coli* CFT073 have highest similarity to SPATES (seriene protease autotransporters of Enterobacteriaceae) proteins made by diarrheagenic *E. coli* and *Shigella species*. Experiment shows SAT possess toxic activity against bladder and kidney cell lines and theraby may have an important role in the pathogenesis of UTI.

VAT was originally discovered in avian pathogenic *E. coli*. In addition, there are Pic and Tsh autotransporters recognized. Pic is known to have seriene protease activity while Tsh lacks. These both are seen to be more prevalent in pyelonephritis strains than fecal strains.

Moreover, recent studies have identified other proteins which are secreted by UPEC and known to have cytotoxic activity. These are NRPS (nonribosomal peptide synthases) and PKS (polyketide synthases) which are produced by B2 *E. coli* strains and are involved in arresting cell cycle [3].

4.3 Iron acquisition by UPEC

Iron, a necessary cofactor for enzymes found in all organisms, remains concealed by iron-binding proteins in humans. Iron is very crucial for the growth of the bacteria. Therefore, bacteria colonizing and causing infections in humans should have some systems to obtain it. Iron is present at a very low concentration at the infection site of the urinary tract and UPEC is known to have multiple systems for iron scavenging. One such potent way to hunt iron is the possession of siderophores by bacteria as it has got a very high affinity for Fe^{3+} that enables *E. coli* to escort iron back to the cell. It is documented that siderophores are usual in *E. coli* strains causing UTI as compared to fecal strains [34]. UPEC encodes a siderophore called enterobactin (which is also present in commensal bacterium) in addition it contains multiple other iron acquisition systems such as yersiniabactin, salmochelin and aerobactin [35]. The gene that encodes aerobactin has been identified to be *iutA*. Furthermore, there is fact that UPEC does not use siderophores alone for scavenging iron, it also employs other iron receptors that are found in the outer membrane which binds iron and takes it back inside the bacterial cell. CFT073 (prototypical UPEC strain) encodes for 14 such outer membrane iron receptors [36].

4.4 Extracellular polysaccharides

There are a variety of extracellular polysaccharides produced by *E. coli* such as O antigen, core polysaccharides of LPS (lipopolysaccharide), colonic acid or capsule, etc. Their role in the pathogenesis of UTI is not well understood. `LPS of UPEC is regarded important in stimulating proinflammatory response in cases of uncomplicated UTI. Also, it is also noted in animal models that acute renal failure due to LPS is not dependant on the presence of functional LPS receptors TLR4 in the kidney but systemic response to LPS.

The capsule is known to provide protection against phagocytosis and complement-mediated bactericidal effect in the host. In addition, it has been found that the K2 capsule and not the K54 capsule acts to be confirmed urovirulence factor [3, 37].

4.5 Proteases

Proteases are enzymes that leads to the cleavage of peptide bonds. They are generally found in mobile genetic elements like transposons, plasmids or prophages [38]. Although, they are not needed for the survival and replication of bacteria, they may be vital for virulence.

Omptins are considered outer-membrane proteases seen in various members of the Enterobacteriaceae family. *E. coli* can encode upto 3 omptims such as *ompT, ompP and arlC* [39]. OmpT have been recognized as a significant virulence factor in UPEC strains causing cystitis, pyelonephritis, urosepsis as opposed to asymptomatic strains [39].

4.6 Flagella

Flagella is an organelle that accounts for the motility in bacteria and flagellated UPEC is responsible for nearly about 70–90% of all UTIs. Genes for synthesis of flagella form a well-regulated and directed cascade of three 3 classes. The benefits of having flagella mediated motility by *E. coli* during colonization of the urinary tract include the capability of dissemination to new sites of the urinary tract to obtain nutrients and in addition to escape from immune responses of the host [4].

4.7 Chemotaxis

Chemotaxis is generally a behavior that is used by the bacteria to sense and then respond to external chemical signals. There are mainly four chemotaxis protein receptors that are necessary for chemotaxis in *Escherichia coli* such as Tar and Tsr (amino acids), Trg (saccharides) and Tap (dipeptides) [40]. There is an observation of *tar* and *tsr* being present in 100% and 98% of all motile UPEC isolates respectively. While *trg* and *tap* were found significantly less among UPEC in comparison to fecal isolates [41].

5. Antimicrobial resistance in UPEC

Antimicrobial therapy is generally recommended for all symptomatic UTI cases including uncomplicated and complicated cases. The choice of an antibiotic should be led by spectrum and susceptibility patterns of the causative agent, its efficacy for the particular indication, tolerability and adverse events, costs and availability, etc. The most commonly used antimicrobials for treating uncomplicated cases include nitrofurantoin, cotrimoxazole (first line), fosfomycin and pivmecillinam (alternative), fluoroquinolones (second line). Ceftriaxone, cefepime, piperacillin-tazobactam, aminoglycosides and carbapenams are used for complicated cases [42]. The international guidelines recommend the use of nitrofurantoin, fosfomycin trometamol and trimethoprim-sulfamethoxazole for the treatment of uncomplicated UTI. Fluoroquinolones in such situations should be kept in reserved as they are used to treat complicated UTI and mild to moderate pyelonephritis [42].

UTIs and especially recurrent UTIs are associated with significant use of antibiotics that promotes resistance. Antimicrobial resistance in UPEC and the spreading of multidrug resistant (MDR) UPEC is a concerning clinical problem, particularly in women with recurrent UTIs. Increasing ineffectiveness of the antimicrobials has led to the emergence of MDR UPEC (resistance to at least one antibiotic in three or more classes) and XDR UPEC (resistance to at least one in all but at least two or fewer classes) [43].

The various mechanisms responsible for resistance include:

1. Bacterial mechanisms

    - Target site inactivation

    - Presence of β-lactamases enzyme

    - Efflux pumps mechanism

    - Through mobile genetic elements such as transposons, gene cassettes, insertions sequences, integrons, etc.

2. Antibiotic consumption without bacterial characterization of the UTI pathogen.

3. Over the counter availability of antimicrobials, thus leading to its overuse.

5.1 Drug resistance mechanisms in different classes of antibiotics

5.1.1 Beta lactam drugs

They are cell wall synthesis inhibitors. One of the main mechanisms of resistance in UPEC is the production of the β-lactamase enzyme which is encoded by the *bla* genes, located on the plasmids. ESBL (extended-spectrum β-lactamase) is one of the types of β-lactamases and is responsible for conferring resistance to penicillin, cephalosporins and monobactams. ESBLs are susceptible to cephamycins, carbapenems and beta lactamase inhibitors; thus making carbapenems drug of choice in such cases. CTX-M, TEM and SHV are most common ESBLs observed among UPEC.

5.1.2 Fluoroquinolones

They are bactericidal drugs. It functions by inhibiting the enzymes topoisomerase II (DNA gyrase, which is encoded by gyr A and gyr B genes) and topoisomerase IV (parC and parE) involved in winding—unwinding of the DNA. It is one of the most widely prescribed drugs due to easily available oral formulations. Fluoroquinolone resistance may be due to chromosomal mutations, plasmid mutations, alteration of outer membrane proteins causing decreased antibiotic uptake and presence of efflux pump systems.

Chromosomal mutation in DNA sequences of gyr A QRDRs i.e., quinolone resistance determining region is predominantly responsible for fluoroquinolone resistance. Resistance is also mediated via PMQR (plasmid-mediated quinolone resistance) genes which include qnr genes such as qnrA, qnr B and qnr C.

5.1.3 Fosfomycin

It is also a bactericidal drug that blocks cell wall synthesis at early stages. It's a widely used drug as a single dose of 3 g is used for the treatment of uncomplicated UTI, and also because of its activity against ESBL and MBL (metallo-β-lactamases). Resistance to fosfomycin is not widespread till date, however, if present may be conferred by:

- Presence of efflux pumps
- Enzymatic cleavage by Fos A, Fos X
- Point mutations

5.1.4 Trimethoprim-sulfamethoxazole

It works by inhibiting the folate synthesis pathway thus inhibiting the DNA synthesis in the susceptible organism. Though widely used as a first-line therapy for uncomplicated cystitis, the use of trimethoprim-sulfamethoxazole has been stopped for empirical use because of its resistance. Over 20% of the UPEC isolates are resistant to trimethoprim-sulfamethoxazole. Resistance to the drug is mediated by

- Efflux pump mechanism
- Target enzyme modification
- Mutational changes in target enzymes

5.1.5 Nitrofurantoin

Reactive intermediates are formed from nitrofurantoin by the action of flavoproteins as a result of which the bacterial ribosomal proteins are inactivated. Resistance to nitrofurantoin till date is much lesser in comparison to the other drugs available, this is attributed to the fact that it acts on multiple targets in the bacterial cell. Resistance may develop due to gene mutations such as nsfA and nfsB genes [44].

6. New therapeutic options, what is in pipeline?

Emerging drug resistance has led to decreased therapeutic options. Therefore, it is imperative to find alternative treatment options. The options are in pipeline for

Newer therapeutic and disease target	**Target in UPEC**	**Available evidence**
Anti-adhesives		
Mannosides for acute cystitis	Type 1 pili (FimH)	In a mouse model, the reduced bacterial burden was observed following treatment and also as a prophylactic agent in mouse models [46, 47]
Galactosides for chronic cystitis and/or pyelonephritis	Fim-like (Fml) pili	In experimental mouse model infected with chronic UTI, reduced bacterial burden in the bladder and kidney [48]

Newer therapeutic and disease target	Target in UPEC	Available evidence
Vaccine		
FimCH against acute and chronic cystitis	UPEC expressing type 1 pili	Phase 1 clinical trial showing no safety concerns and a reduction in total UTI recurrence in treatment cohort; approved for compassionate use as an investigational intervention
Uro-vaxom	Contains a lyophilized mix of membrane proteins from 18 different strains of *E. coli*	Licensed in 30 countries, represents a safe and effective treatment option for prophylaxis of recurrent UTIs [49]
Urovac		Recurrent UTI [49]
Others		
Non-steroidal anti-inflammatory drugs (NSAIDS)—chronic and recurrent cystitis	UPEC	*In vivo* mouse models. It demonstrates reduced bladder remodeling, and in human clinical studies demonstrate effective resolution of symptoms with a reduction in overall antibiotic used when used in place of antibiotics [50]
HIF-1α inhibition (AKB-4924)	UPEC	Decreased adherence and invasion of UPEC of cultured human uroepithelial cells, decreased inflammation and bacterial load in mouse models of infection [51]
Probiotics—recurrent UTI	UPEC	Reduced frequency of infection in various patient populations having rUTI [52]
Lactoferrin—cystitis	UPEC	Reduced adherence to human bladder epithelial cell lines and reduced mouse bladder bacterial burdens following treatment with exogenous lactoferrin [53]
Antibiotic—recarbrio (Imipenem, cilastatin and relebactam)		Approved by FDA (U.S. Food and Drug Administration) for the treatment of complicated UTI

Table 3.
Newer therapeutics options in pipeline for the treatment of urinary tract infections caused by UPEC [45].

preventing and treating infections caused by UPEC includes antibiotic recarbrio (FDA approved), vaccines, anti-adhesives etc. **Table 3** highlights newer therapeutics options in the pipeline for the treatment of urinary tract infections caused by UPEC [54].

7. Conclusion

Urinary tract infections, community and hospital-associated affects millions of people especially women worldwide each year. As UPEC is the most common microorganism behind these infections, so adequate knowledge and understanding of UPEC, its virulence factors and antimicrobial resistance pattern is vital for adequate treatment and prevent recurrences. In addition, this can help in developing new therapeutics options in the era of widespread antimicrobial resistance.

Conflict of interest

None.

Author details

Navneet Kaur[1], Ashwini Agarwal[1*], Malika Grover[2] and Sanampreet Singh[3]

1 Department of Microbiology, All India Institute of Medical Sciences, Rajkot, Gujarat, India

2 Department of Microbiology, All India Institute of Medical Sciences, Jodhpur, Rajasthan, India

3 Independent Researcher, Newport Beach, California, United States of America

*Address all correspondence to: ash.afmc@gmail.com

References

[1] Kucheria R, Dasgupta P, Sacks S, K han M, Sheerin N. Urinary tract infections: New insights into a common problem. Postgraduate Medical Journal. 2005;**952**:83. DOI: 10.1136/pgmj.2004.023036

[2] Loh KY, Sivalingam N. Urinary tract infections in pregnancy. Malaysian Family Physician. 2007;**2**:54-57

[3] Mobley HL, Donnenberg MS, Hagan EC. Uropathogenic *Escherichia coli*. EcoSal Plus. 2009;**3**(2):1-27. DOI: 10.1128/ecosalplus.8.6.1.3

[4] Bien J, Sokolova O, Bozko P. Role of uropathogenic *Escherichia coli* virulence factors in development of urinary tract infection and kidney damage. International Journal of Nephrology. 2012;**2012**:1-9. DOI: 10.1155/2012/681473

[5] Johnson JR, Kuskowski MA, Gajewski A, Soto S, Horcajada JP, De Anta MT, et al. Extended virulence genotypes and phylogenetic background of *Escherichia coli* isolates from patients with cystitis, pyelonephritis, or prostatitis. The Journal of Infectious Diseases. 2005;**191**:46-50. DOI: 10.1086/426450

[6] Ballesteros-Monrreal MG, Arenas-Hernández MM, Enciso-Martínez Y, Martínez-de la Peña CF, Rocha-Gracia RDC, Lozano-Zaraín P, et al. Virulence and resistance determinants of uropathogenic *Escherichia coli* strains isolated from pregnant and non-pregnant women from two states in mexico. Infection and Drug Resistance. 2020;**13**:295. DOI: 10.2147/IDR.S226215

[7] Mobley HL, Green DM, Trifillis AL, Johnson DE, Chippendale GR, Lockatell CV, et al. Pyelonephritogenic *Escherichia coli* and killing of cultured human renal proximal tubular epithelial cells: Role of hemolysin in some strains. Infection and Immunity. 1990;**58**:1281-1289. DOI: 10.1128/iai.58.5.1281-1289.1990

[8] Luo C, Hu GQ, Zhu H. Genome reannotation of *Escherichia coli* CFT073 with new insights into virulence. BMC Genomics. 2009;**10**:1. DOI: 10.1186/1471-2164-10-552

[9] Emödy L, Kerényi M, Nagy G. Virulence factors of uropathogenic *Escherichia coli*. International Journal of Antimicrobial Agents. 2003;**22**:S29-S33. DOI: 10.1016/S0924-8579(03)00236-X

[10] Edén CS, Jodal U, Hanson LA, Lindberg U, Åkerlund AS. Variable adherence to normal human urinary-tract epithelial cells of *Escherichia coli* strains associated with various forms of urinary-tract infection. The Lancet. 1976;**308**:490-492. DOI: 10.1016/S0140-6736(76)90788-1

[11] Korhonen TK, Virkola R, Holthöfer H. Localization of binding sites for purified *Escherichia coli* P fimbriae in the human kidney. Infection and Immunity. 1986;**54**:328-332. DOI: 10.1128/iai.54.2.328-332.1986

[12] Plos K, Connell H, Jodal U, Marklund BI, Mårild S, Wettergren B, et al. Intestinal carriage of P fimbriated *Escherichia coli* and the susceptibility to urinary tract infection in young children. Journal of Infectious Diseases. 1995;**171**:625-631. DOI: 10.1093/infdis/171.3.625

[13] Wullt B, Bergsten G, Samuelsson M, Gebretsadik N, Hull R, Svanborg C. The role of P fimbriae for colonization and

host response induction in the human urinary tract. The Journal of Infectious Diseases. 2001;**183**:s43-s46

[14] Otto G, Magnusson M, Svensson M, Braconier J, Svanborg C. pap genotype and P fimbrial expression in *Escherichia coli* causing bacteremic and nonbacteremic febrile urinary tract infection. Clinical Infectious Diseases. 2001;**32**:1523-1531. DOI: 10.1086/320511

[15] Plos K, Lomberg H, Hull S, Johansson I, Svanborg C. *Escherichia coli* in patients with renal scarring: Genotype and phenotype of Galα1-4Galβ-, Forssman- and mannose-specific adhesins. Pediatric Infectious Disease Journal. 1991;**10**:15-19

[16] Zhang L, Foxman B. Molecular epidemiology of *Escherichia coli* mediated urinary tract infections. Frontiers in Bioscience. 2003;**8**:e235-e244

[17] Oelschlaeger TA, Dobrindt U, Hacker J. Virulence factors of uropathogens. Current Opinion in Urology. 2002;**12**:33-38. DOI: 10.1097/00042307-200201000-00007

[18] Johnson JR. Microbial virulence determinants and the pathogenesis of urinary tract infection. Infectious Disease Clinics. 2003;**17**:261-278. DOI: 10.1016/S0891-5520(03)00027-8

[19] Marrs CF, Zhang L, Tallman P, Manning SD, Somsel P, Raz P, et al. Variations in 10 putative uropathogen virulence genes among urinary, faecal and peri-urethral *Escherichia coli*. Journal of Medical Microbiology. 2002;**51**:138-142

[20] Marrs CF, Zhang L, Foxman B. *Escherichia coli* mediated urinary tract infections: Are there distinct uropathogenic *E. coli* (UPEC) pathotypes? FEMS Microbiology Letters. 2005;**252**:183-190. DOI: 10.1016/j.femsle.2005.08.028

[21] O'Brien GJ, Chambers ST, Peddie B, Mahanty KH. The association between colicinogenicity and pathogenesis among uropathogenic isolates of *Escherichia coli*. Microbial Pathogenesis. 1996;**20**(3):185-190. DOI: 10.1006/mpat.1996.0017

[22] Kostakioti M, Hultgren SJ, Hadjifrangiskou M. Molecular blueprint of uropathogenic Escherichia coli virulence provides clues toward the development of anti-virulence therapeutics. Virulence. 2012;**3**(7):592-593

[23] Nowicki B, Labigne A, Moseley S, Hull R, Hull S, Moulds J. The Dr hemagglutinin, afimbrial adhesins AFA-I and AFA-III, and F1845 fimbriae of uropathogenic and diarrhea-associated *Escherichia coli* belong to a family of hemagglutinins with Dr receptor recognition. Infection and Immunity. 1990;**58**(1):279-281. DOI: 10.1128/iai.58.1.279-281.1990

[24] Das M, Hart-Van Tassell A, Urvil PT, Lea S, Pettigrew D, Anderson KL, et al. Hydrophilic domain II of *Escherichia coli* Dr fimbriae facilitates cell invasion. Infection and Immunity. 2005;**73**:6119-6126. DOI: 10.1128/IAI.73.9.6119-6126.2005

[25] Hacker JS, Morschhäuser J. In: Klemm P, editor. Fimbriae, Adhesion, Genetics, Biogenesis, and Vaccines. Boca Raton, Fla, USA: CRC Press; 1994. pp. 27-36

[26] Marre R, Kreft B, Hacker J. Genetically engineered S and F1C fimbriae differ in their contribution to adherence of *Escherichia coli* to cultured renal tubular cells. Infection and Immunity. 1990;**58**:3434-3437. DOI: 10.1128/iai.58.10.3434-3437.1990

[27] Donnenberg MS, Welch RA. In: Mobley HLT, Warren JW, editors. Virulence Determinants of Uropathogenic *Escherichia coli* Urinary Tract Infections: Molecular Pathogenesis and Clinical Management. Washington, D.C: American Society for Microbiology; 1996. pp. 135-174

[28] Foxman B, Barlow R, D'Arcy H, Gillespie B, Sobel JD. Urinary tract infection: Self-reported incidence and associated costs. Annals of Epidemiology. 2000;**10**:509-515. DOI: 10.1016/S1047-2797(00)00072-7

[29] Dudgeon L, Worldley E, Bawtree F. On *Bacillus coli* infections of the urinary tract, especially in relation to hemolytic organisms. Journal of Hygiene. 1921;**10**:137-164. DOI: 10.1017/S002217240003391X

[30] Chen M, Tofighi R, Bao W, et al. Carbon monoxide prevents apoptosis induced by uropathogenic *Escherichia coli* toxins. Pediatric Nephrology. 2006;**21**:382-389. DOI: 10.1007/s00467-005-2140-1

[31] Jakobsson B, Berg U, Svensson L. Renal scarring after acute pyelonephritis. Archives of Disease in Childhood. 1994;**70**:111-115. DOI: 10.1136/adc.70.2.111

[32] Mills M, Meysick KC, O'Brien AD. Cytotoxic necrotizing factor type 1 of uropathogenic *Escherichia coli* kills cultured human uroepithelial 5637 cells by an apoptotic mechanism. Infection and Immunity. 2000;**68**:5869-5880. DOI: 10.1128/IAI.68.10.5869-5880.2000

[33] Parham NJ, Pollard SJ, Desvaux M, Scott-Tucker A, Liu C, Fivian A, Henderson IR. Distribution of the serine protease autotransporters of the Enterobacteriaceae among extraintestinal clinical isolates of *Escherichia coli*. Journal of Clinical Microbiology. 2005;**43**(8):4076-4082. DOI: 10.1128/ JCM.43.8.4076-4082.2005

[34] Johnson JR, Moseley SL, Roberts PL, Stamm WE. Aerobactin and other virulence factor genes among strains of *Escherichia coli* causing urosepsis: Association with patient characteristics. Infection and Immunity. 1988;**56**:405-412. DOI: 10.1128/iai.56.2.405-412.1988

[35] Garcia EC, Brumbaugh AR, Mobley HL. Redundancy and specificity of *Escherichia coli* iron acquisition systems during urinary tract infection. Infection and Immunity. 2011;**79**(3):1225-1235. DOI: 10.1128/IAI.01222-10

[36] Welch RA, Burland V, Plunkett GI, Redford P, Roesch P, Rasko D, et al. Extensive mosaic structure revealed by the complete genome sequence of uropathogenic *Escherichia coli*. Proceedings of the National Academy of Sciences. 2002;**99**:17020-17024. DOI: 10.1073/pnas.252529799

[37] Cunningham PN, Wang Y, Guo R, He G, Quigg RJ. Role of Toll-like receptor 4 in endotoxin-induced acute renal failure. Journal of Immunology. 2004;**172**:2629-2635. DOI: 10.4049/jimmunol.172.4.2629

[38] Potempa J, Pike RN. Bacterial peptidases. Contributions to Microbiology. 2005;**12**:132-180. DOI: 10.1080/14756366.2021.1937619

[39] Lundrigan MD, Webb RM. Prevalence of ompT among *Escherichia coli* isolates of human origin. FEMS Microbiology Letters. 1992;**97**:51-56. DOI: 10.1111/j.1574-6968.1992.tb05438.x

[40] Hedblom ML, Adler J. Genetic and biochemical properties of *Escherichia coli* mutants with defects in serine

chemotaxis. Journal of Bacteriology. 1980;**144**:1048-1060. DOI: 10.1128/jb.144.3.1048-1060.1980

[41] Lane MC, Lloyd AL, Markyvech TA, Hagan EC, Mobley HL. Uropathogenic *Escherichia coli* strains generally lack functional Trg and Tap chemoreceptors found in the majority of *E. coli* strains strictly residing in the gut. Journal of Bacteriology. 2006;**188**:5618-5625. DOI: 10.1128/JB.00449-06

[42] European Association of Urology. Urological infections. Neitherlands: EAU; 2020

[43] Ochoa SA, Cruz-Córdova A, Luna-Pineda VM, Reyes-Grajeda JP, Cázares-Domínguez V, Escalona G, et al. Multidrug-and extensively drug-resistant uropathogenic *Escherichia coli* clinical strains: Phylogenetic groups widely associated with integrons maintain high genetic diversity. Frontiers in Microbiology. 2016;7:2042. DOI: 10.3389/fmicb.2016.02042

[44] Kot B. Antibiotic resistance among uropathogenic *Escherichia coli*. Polish Journal of Microbiology. 2019;**68**:403. DOI: 10.33073/pjm-2019-048

[45] Klein RD, Hultgren SJ. Urinary tract infections: Microbial pathogenesis, host–pathogen interactions and new treatment strategies. Nature Reviews Microbiology. 2020;**18**:211-226. DOI: 10.1038/s41579-020-0324-0

[46] Jarvis C et al. Antivirulence isoquinolone mannosides: Optimization of the biaryl aglycone for FimH lectin binding affinity and efficacy in the treatment of chronic UTI. ChemMedChem. 2016;**11**:367-373. DOI: 10.1002/cmdc.201600006

[47] Mydock-McGrane L et al. Antivirulence C-mannosides as antibiotic-sparing, oral therapeutics for urinary tract infections. Journal of Medicinal Chemistry. 2016;**59**:9390-9408. DOI: 10.1021/acs.jmedchem.6b00948

[48] Kalas V, Hibbing ME, Maddirala AR, Chugani R, Pinkner JS, Mydock-McGrane LK, et al. Structure-based discovery of glycomimetic FmlH ligands as inhibitors of bacterial adhesion during urinary tract infection. Proceedings of the National Academy of Sciences. 2018;**115**:E2819-E2828. DOI: 10.1073/pnas.1720140115

[49] Aziminia N, Hadjipavlou M, Philippou Y, Pandian SS, Malde S, Hammadeh MY. Vaccines for the prevention of recurrent urinary tract infections: A systematic review. BJU International. 2019;**123**:753-768. DOI: 10.1111/bju.14606

[50] Bleidorn J, Gágyor I, Kochen MM, Wegscheider K, Hummers-Pradier E. Symptomatic treatment (ibuprofen) or antibiotics (ciprofloxacin) for uncomplicated urinary tract infection?-results of a randomized controlled pilot trial. BMC Medicine. 2010;**8**:1-8. DOI: 10.1186/1741-7015-8-3

[51] Lin AE, Beasley FC, Olson J, Keller N, Shalwitz RA, Hannan TJ, et al. Role of hypoxia inducible factor-1α (HIF-1α) in innate defense against uropathogenic *Escherichia coli* infection. PLoS Pathogens. 2015;**11**:e1004818. DOI: 10.1371/journal.ppat.1004818

[52] Darouiche RO, Green BG, Donovan WH, Chen D, Schwartz M, Merritt J, et al. Multicenter randomized controlled trial of bacterial interference for prevention of urinary tract infection in patients with neurogenic bladder. Urology. 2011;**78**:341-346. DOI: 10.1016/j.urology.2011.03.062

[53] Patras KA, Ha AD, Rooholfada E, Olson J, Rao SP, Lin AE, et al. Augmentation of urinary lactoferrin enhances host innate immune clearance of uropathogenic *Escherichia coli*. Journal of Innate Immunity. 2019;**11**(6):481-495. DOI: 10.1159/000499342

[54] Terlizzi ME, Gribaudo G, Maffei ME. UroPathogenic *Escherichia coli* (UPEC) infections: Virulence factors, bladder responses, antibiotic, and non-antibiotic antimicrobial strategies. Frontiers in Microbiology. 2017;**8**:1566. DOI: 10.3389/fmicb.2017.01566

Chapter 2

Bacteriophage Therapy for Urinary Tract Infections Caused by *Escherichia coli*

Sonia Bhonchal Bhardwaj

Abstract

Urinary tract infections (UTIs) are the most prevalent bacterial diseases affecting 150 million people annually worldwide. Around 85% of UTIs are caused by *Escherichia coli* from the Enterobacteriaceae family. The pathogenesis of uropathogenic *E. coli* (UPEC) involves adherence, colonization, evading host defenses, and damage to host tissue to achieve virulence. The uncontrolled use of antibiotics worldwide during therapy of UTIs has resulted in increased antibiotic resistance and the emergence of multidrug resistance (MDR) and extensive drug resistant (XDR) to UPEC. Bacteriophages have the potential to eliminate and manage resistant biofilm-forming uropathogenic organisms, such as *E. coli* and control UTIs. The chapter discusses the use of phages as an alternative treatment for UTIs caused by UPEC.

Keywords: urinary tract infection, uropathogenic *E. coli*, bacteriophages

1. Introduction

Urinary tract infections (UTIs) are one of the most frequent bacterial infections and the primary causative agent is *Escherichia coli* [1]. The other causative agents reported for UTIs include *Staphylococcus saprophyticus*, *Pseudomonas aeruginosa,* Klebsiella, Enterobacter, and Proteus species [2]. *E. coli*, the primary causative agent of UTI, is a Gram-negative bacteria from the Enterobacteriaceae family. The enteropathogenic strains of *E. coli* are divided into two types: intestinal *E. coli,* which have enteropathogenic E. coli (EPEC), enterohemorrhagic *E. coli* (EHEC), enteroinvasive *E. coli*, enteroaggregative *E. coli*, enterotoxigenic *E. coli* (ETEC). The second category is the uropathogenic *E. coli,* causing extraintestinal infections (**Table 1**). The uropathogenic serotypes 01 0K1, 06:K2, 04:K12, 016:K1, or 018:K5 are associated with the majority of UTIs [3]. Uropathogenic *E. coli* causes both complicated and uncomplicated UTIs. UPEC has fimbria as an important virulence factor. These fimbriae contain subunit protein (pap A) encoded by papA gene [4]. The type I fimbriae are most commonly expressed adhesins that allow the bacteria to attach and colonize the human urogenital tract. The type I fimbriae constitutes of Fim H protein (responsible for binding), Iaminin (part of extracellular matrix), and secretory Ig A. Another fimbriae present in UPEC is P fimbriae (PapG) adhesion of p fimbriae mediates the bacterial binding, thus inducing UTI symptoms [4]. Other

Diarrheagenic *E. coli* or enteric *E. coli*	Extraintestinal *E. coli* (EXPEC)
Pathotypes. 1. Enterotoigenic E .coli (ETEC). 2. Enteropathogenic *E. coli* (EPEC). 3. Shiga-toxin-producing *E. coli* (STEC). 4. Enteroaggregative *E. coli* (EAEC). 5. Enteroinvasive *E. coli* (EIEC). 6. Diffusely adherent *E. coli* (DAEC).	Pathotypes: uropathogenic *E. coli*, neonatal meningitis *E. coli* (NMEC) Serotypes.O1:H4, O1:H6,O1:H7,O1:H, O2:H1, O2:H4, O4:H5,O6:H1, O7:H4,O7:H6, O7:H-, O18ac:H7, O18ac:H-, O22:H, O25:H1, O75:H5 and O75:H7.

Table 1.
Pathogenic E. coli.

virulence factors in UPEC are F1C and S fimbriae enclosing the fac and sfa gene, cytotoxic necrotic zing factor (NF1), iron-binding siderophores, and K1 capsular polysaccharide.

Bacteriophages are viruses that attack bacteria. Antimicrobial resistance by bacteria has now become a global threat and could kill 50 million people by the year 2050 as per the World Health Organization estimates [5]. Phages are now known to cure antibiotic-resistant bacterial infections as well as decrease bacterial virulence by overcoming the barriers bacteria used to avoid them. Bacteriophages are now being explored as potential therapeutic tools for the elimination of bacterial pathogens. Bacteriophages can disrupt pathogenic processes associated with biofilm and exopolysaccharide formation by microflora. Bacteriophage therapy is a promising strategy to control bacterial infections as phages are very efficient in killing host bacteria and do not disrupt other flora and have a low cost of production [6]. Antibiotics used alone have a broad spectrum of activity inducing drug resistance in bacteria and are toxic, whereas phages are non-toxic [7]. Using other strategies, such as herbal products, is more costly and less efficient and has a broad spectrum when compared with phages that are safe and efficient even through oral administration [8]. The combination of antibiotics and phage therapy and the use of phage cocktails have great potential in the treatment of drug-resistant bacterial infections, particularly UTIs. This chapter focuses on the use of phages in treating UTIs caused by uropathogenic *E. coli*.

2. Phages as therapeutics for uropathogenic *E. coli* (UPEC)

Newer therapeutic options like phages are alternative treatment options for treating UTIs. Phages have been tried as a potential candidate for treating UTIs in a number of studies. Six lytic bacteriophages each at a titer of 10^6 p.f.u/ml to *P. aeruginosa* causing UTI showed a decrease in the number of target bacteria [9]. In another study, Pyo bacteriophages to *S. aureus*, *E. coli*, Streptococcus species, *P. aeruginosa*, and Proteus species causing UTI were given twice to patients after transurethral resection of the prostate in a solution form. The patients were asked to retain the phage solution for 30–60 minutes in their bladders. After phage therapy, a decreased bacterial count was seen in 67% of patients with no side effects [10]. In a recent study by Lorenz et al. a pyobacteriophage cocktail solution was given twice daily for 7 days in UTI and was found to be comparable to regular antibiotic therapy [11, 12]. Emerging antimicrobial resistance in UPEC has led to the emergence of multidrug-resistant UPEC and extensively drug-resistant UPEC [10]. Initial studies showed the phage preparations were locally and orally applied to UTIs caused by *E. coli*, Staphylococcus, and Proteus species. A bacterial reduction of 84% was observed [13].

Broadly bacteriophages are now being used in infecting UPEC strains causing UTIs in four ways (**Table 2**).

a. Phage cocktails

b. Genetically engineered phages

c. Phage lytic proteins

d. Phages in combination with antibiotics

2.1 Phage cocktails against UPEC

Monophage therapy or using a single phage has an important limitation, which is a narrow host range. Phage cocktails use two or more phages for therapy making the host range broad and overcoming host bacterial resistance to phages. The phage cocktail or combination of phages to UPEC can recognize more than one host receptor and hence infect many uropathogenic strains. A phage cocktail of T4 phage and KEP10 phage was introduced in the peritoneal cavity of the mouse as the first therapeutic candidate for the treatment of UTI caused by UPEC [14]. The efficacy of T1, T4, and phiX174-like phages was also evaluated against UPEC. T1 phage was found to be the most effective in killing UPEC as it had a broad lytic spectrum; however, a combination of T1, T4, and phiX174 was capable of infecting a variety of antibiotic-tolerant UPEC strains [15]. A cocktail of nine phages without horizontal gene transfer and undesired genes from 99 T4-like coliphages to UPEC was used to produce a cocktail and given to 15 healthy adults. No side effects were seen, indicating that phage therapy was safe to use in UTIs [16]. In a study, it was seen that phage SP21 uses OmpC of *E. coli* 0157:H7 as a receptor, when this receptor was deleted, the phage-resistant bacteria emerged after 8 hrs of incubation. On modifying the lipopolysaccharide of the bacteria, the resistant bacteria emerged after 6 hrs of incubation with phage SP22. When a combination of two phages SP21 and SP22 binding to different host receptors of EHEC (*E. coli* 0157:H7) was used, it resulted in significant delay at the time of emergence of phage-resistant *E. coli* (upto 30 hrs) as compared to phages used alone [29]. Dual receptor phages to UPEC, which identify more than one receptor, have also been identified. Dual receptor phages to UPEC reported are T4 phages, T2 phages, and phage K1-5 of the family Podaviridae that infects both K1 and K5 strains of *E. coli* [17, 30, 31]. UPEC causes UTI by adhering to the urothelium producing biofilms successfully evading them from the host immune system and antibiotics. Phage cocktails have been found to be suitable for killing bacteria in biofilms. Biofilms of *E. coli* on the surface of polyvinyl chloride were susceptible to phage T4D+ [32]. Phage cocktails can be used for treating UTIs caused by *E. coli* biofilms present on urinary catheters. However, any mutational or conformational change in the host bacterial receptors can make the phages resistant to the bacteria, which is a limitation of using this strategy.

(1) Phage cocktails	a. phage cocktail of T4 and KEP 10 phage induced in the peritoneal cavity of mouse for treatment of UTI caused by UPEC [14]. b. Phage cocktail of T1, T4 and phi X 174 like phages evaluated against UPEC were capable of lysing a variety of UPEC strains [15]. c. Phage cocktail of 9 phages from 99 T4 like coliphages to UPEC was found to be safe and effective in UTI [16]. d. Dual receptor phages T4, T2 and K1-5 infecting K1 and K5 strains of UPEC were found to be efficient [17].
(2) Genetically engineered phages	a. Enzymatic engineered phage T7DspB was found to be more efficient in reducing biofilm formed by clinical *E. coli* isolates as compared to natural lytic phage T7 [18]. b. Engineered phage K1F-GFP was found to be very effective in killing host bacteria *E. coli* present in T24 epithelial cells of human urinary bladder [19]. c. Genetically engineered ε2 phages were found to be more effective for 47 *E. coli* strains found in UTI.
(3) Phage lytic proteins	a. *E. coli* specific phage lyase lysep 3 fused with N-terminal region of Bacillus amyloliquefaciens was found to be highly efficient in lysing clinical isolates of *E. coli* [20]. b. Phage lytic proteins in combination with chelating agent like EDTA was used to disrupt Gram-bacterial cell outer membrane barrier [21]. c. Endolysins called 'artilysins ' which can distort the LPS and has high antibacterial effect against isolates of *E. coli* [22]. Artilysin Art-175 had high bactericidal activity against colistin resistant *E. coli* isolates [23]. d. Phage lysin $LySep^3$ has high increased antibacterial activity against *E. coli* [24]. e. Innolysins [combination of fused phage T5 endolysin and phage receptor binding proteins (RBPs)]. Innolysin Ec6 and Ec21 was found to be highly effective against UPEC [25].
(4) Phages in combination with antibiotics	a. T4 phage and cefotaxime were highly effective in destruction of T4 host *E. coli* ATCC11303 biofilms as compared to antibiotic given alone [26]. b. T4 phage with beta lactam, quinolone and mitomycin C were more effective in destruction of *E. coli* biofilms [27]. c. Phage cocktail with antibiotics was found to be effective in combating drug resistant uropathogens [28].

Table 2.
Types of phage therapy for UPEC.

2.2 Genetically engineered phages against UPEC

Genetically modified or engineered phages have been reported for use in UTIs particularly multidrug-resistant uropathogens. These genetically engineered phages having desirable properties are made using genetic engineering methods, such as homologous recombination, phage recombination of electroporated DNA, *in vivo* recombination, and CRISPR-CAS- mediated genome engineering [33]. An enzymatic engineered phage T7DspB, which expresses exopolysaccharide (EPS)-degrading enzyme dispersin B (DspB), hydrolyses an adhesin required by *E. coli* K12 and clinical *E. coli* isolates for biofilm formation. This genetically modified phage T7DspB had more efficiency in reducing biofilm as compared to natural lytic phage T7 [18]. A phage specific for UPEC (*E. coli* K1) has been genetically modified using the CRISPR-CAS mechanism. The phage called K1F-GFP was very effective in killing host bacteria *E. coli* EV 36-RFP present in T24 epithelial cells of the human urinary bladder [19]. A recent study shows ε^2 phages having mosaic intercrossing of 2–3 ancestor phages and devoid of genes conferring lysogeny, antibiotic resistance, or virulence were more virulent and effective for 47 *E. coli* strains found in UTI [34]. Genetically engineered phages can be especially beneficial in the treatment of UTIs caused by multidrug-resistant bacteria, however, the cost factor, narrow host range, and host immune responses are the limitation. The above studies show that engineered phages can be used in killing biofilm-forming *E. coli* causing UTI as future therapy in humans.

2.3 Phage lytic proteins for UPEC

With the advancement of genomics phage, lytic proteins or enzymes are being developed. They have high antibacterial activity against biofilm-forming multidrug-resistant clinical isolates. Phages produce cell wall lytic proteins, such as endolysins and virion-associated peptidoglycan hydrolases (PGH). Endolysins or lysins are produced by the phages in the later stages of the lytic cycle. They lyse the host bacteria "from within" when the phage lytic cycle ends [35]. Endolysin integrated with outer membrane permeabilizers (omps) against UPEC and other Gram-negative bacteria, which lead to the lysis of the bacterial cell wall. This endolysin showed high antibacterial activity against the multidrug clinical isolates of Gram-negative bacteria [36]. A study used *E. coli*-specific phage lyase lysep3 fused with the N-terminal region of Bacillus amyloliquefaciens found to be highly efficient in lysing clinical isolates of *E. coli*, *P. aeruginosa*, *A. baumanni*, and Streptococcus strains [20].

Virion-associated peptidoglycan hydrolases (PGH) produced by phages are enzymes that cause "lysis of cell wall from without" thereby killing the host bacteria [37]. Early studies showed the use of phage lytic proteins in combination with a chelating agent like ethylene diamine tetra acetic acid disodium dehydrate (EDTA) to disrupt the Gram-negative bacterial cell outer membrane barrier [21]. Protein engineering techniques are now being used to increase the efficiency of endolysin penetration in UPEC. These endolysins engineered to fuse with OMPs can distort the LPS of the Gram-negative bacteria and are called "artilysins." The first study on artilysin used modular endolysin OBPgp279 of *P. fluorescens* phage and PVP-SE1gp146 of *Salmonella enterica serovar enteridis* phage PVP-SE1 in integration with seven outer membrane peptides. These resulting artilysins had a high antibacterial effect against isolates of *E. coli* [22]. Another artilysin Art-175 was made and tested on colistin-resistant *E. coli* isolates. High bactericidal activity was observed against colistin-resistant *E. coli* isolates [23]. The c-terminal of *E. coli* phage lysin Lysep^3 was genetically engineered. It showed increased antibacterial activity against *E. coli* [24]. Endolysins have recently been engineered as "Innolysins," which combine the binding capacity of phage receptor binding proteins (RBPs). Twelve innolysins were made by fusing phage T5 endolysin and RBPb5 in different configurations. Innolysin Ec6 was highly effective against *E. coli*, innolysin Ec21 displayed bactericidal activity to *E. coli* resistant to third-generation cephalosporins [25].

2.4 Phages in combination with antibiotics

Phage-antibiotic combinations are based on phage-antibiotic synergy (PAS) that is antibiotics are more effective in treating biofilm infections in sub-lethal concentrations combined with phages than phages applied alone. The PAS also significantly reduces the development of bacterial resistance as compared to phages used singly [26]. The first study using phage-antibiotic combination to control *E. coli* biofilm *in vitro* was when T4 phage and cefotaxime resulted in effective destruction of T4 host E. coli ATCC 11303 biofilms as compared when antibiotic was given alone [27]. With other antibiotics, such as beta-lactam, quinolone, and mitomycin C, there was a similar effect and an increase in T4 phage plaque size. PAS has been studied in other pathogens like biofilm-forming Pseudomonas aeruginosa. When *P. aeruginosa* biofilms were treated in combination with phages and different antibiotics like ceftazidime, ciprofloxacin, colistin, gentamicin, and tobramycin showed high bactericidal activity to *P. aeruginosa* in biofilms grown on human epithelial cell culture [38]. Another study has shown the synergism effect of Cpl-711 endolysin of *S. pneumoniae* and amoxicillin or cefixime on multidrug-resistant isolates of *S. pneumoniae* using mouse and zebrafish models for experimental *in vivo* infection [39]. In a recent study, phage cocktail and antibiotics were used together to combat drug-resistant uropathogens (UPEC). Synergistic effects of the phage cocktail with antibiotics showed phage antibiotic synergism at a lower MIC value of antibiotics [28]. The PAS is quite complex and influenced by many factors like phage and class of antibiotics used, at what concentration the phage lowers the MIC value of antibiotics, and the combination is effective on drug-resistant uropathogens besides host factors like urine and serum. Thus, more studies are needed in PAS to make it a successful therapy for uropathogens mainly UPEC. The use of phages and limiting bacteria to nutrients like iron, which have an important role in biofilm development has also been reported. By adding divalent metal ions, such as Co(II) and Zn(II) to the culture medium a reduction in biofilm development by UPEC was seen [40].

3. Conclusion

The major cause of UTIs worldwide is uropathogenic *E. coli*. The development of resistance in uropathogenic *E. coli* is a serious therapeutic problem that requires newer antibiotics and alternative forms of therapy, such as phages. In the treatment of UTIs, studies are being conducted on various forms of bacteriophages, such as phage cocktails, genetically modified phages, phage lytic enzymes and their derivatives, and phage-antibiotic combinations. Clinical trials are being conducted on phage cocktails and phage lytic enzymes for treating UTIs and no randomized control trials. The phage therapy still requires validated clinical research to use different types of phage therapy to eliminate UPEC and the biofilm formed in the urinary tract to control UTIs. More research on phage therapy is still required on drug-resistant uropathogens. Undoubtedly in the future phages can emerge as pharmaceutical compounds, an alternative to conventional antibiotics particularly for treating UTIs caused by drug-resistant uropathogenic *E. coli*.

Conflict of interest

The author has no conflicts of interest.

Author details

Sonia Bhonchal Bhardwaj
Department of Microbiology, Dr. Harvansh Singh Judge Institute of Dental Sciences and Hospital Panjab University, Chandigarh, India

*Address all correspondence to: sbbhardwaj2002@yahoo.com

References

[1] Sahm DF, Thornsberry C, Mayfield DC, Jones ME. Multidrug-resistant urinary tract isolates of *E. coli*: Prevalence and patient demographics in the United States in 2000. Antimicrobial Agents and Chemotherapy. 2001;**45**:1402-1406

[2] Kline KA, Bowdish DM. Infection in an aging population. Current Opinion in Microbiology. 2016;**29**:63-67

[3] Vaisanen-Rhen V, Elo J, Vaisanen E, et al. P fimbriated clones among uropathogenic *E. coli* strains. Infection and Immunity. 1984;**43**:149-155

[4] Li X, Zhou K, Wang J, Guo J, CaO Y, et al. Diagnostic value of the fimbriae distribution pattern in localization of UTI. Frontiers in Medicine. 2021;**18**:602

[5] New report calls for urgent action to avert antimicrobial resistance crisis. 2019. Available from: https://www.who.int/news/item/29-04-2019-new-report-calls-for-urgent-action-to-avert-antimicrobial-resistance-crisis

[6] Sulakvelidze A, Alavidze Z, Morris JG Jr. Bacteriophage therapy. Antimicrobial Agents and Chemotherapy. 2001;**45**:649-659

[7] Sarker SA, Berger B, Deng Y, Kieser S, Huq S, Bardhan PK, et al. Oral application of *E. coli* bacteriophage: Safety tests in healthy and diarrheal children from Bangladesh. Environmental Microbiology. 2017;**19**:237-250

[8] Gindin M, Febre HP, Rao S, Wallace TC, Weir TL. Bacteriophage for gastrointestinal tract (PHAGE) study. Evaluating the safety and tolerability of supplemental bacteriophage consumption. Journal of the American College of Nutrition. 2018;**29**:1-8

[9] Khawaldeh A, Morales S, Dillon B, Alavidze Z, Ginn AN, Thomas L, et al. Bacteriophage therapy for refractory *Pseudomonas aeruginosa* urinary tract infection. Journal of Medical Microbiology. 2011;**60**(11):1697-1700

[10] Ujmajuridze A, Chanishvili N, Goderdzishvili M, Leitner L, Mehnert U, Chkhotua A, et al. Adapted bacteriophages for treating urinary tract infections. Frontiers in Microbiology. 2018;**9**:1832

[11] Leitner L, Ujmajuridze A. Intravesical bacteriophages for treating urinary tract infections in patients undergoing transurethral resection of the prostate: A randomised, placebo-controlled, double-blind clinical trial. The Lancet Infectious Diseases. 2021;**20**:1263-1272

[12] Ochoa SA, Cruz-Córdova A, Luna-Pineda VM, Reyes-Grajeda JP, Cázares-Domínguez V, Escalona G, et al. Multidrug-and extensively drug-resistant uropathogenic *Escherichia coli* clinical strains: Phylogenetic groups widely associated with integrons maintain high genetic diversity. Frontiers in Microbiology. 2016;**21**(7):2042

[13] Perepanova TS, Darbeeva OS, Kotliarova GA, Kondrateva EM, Maiskai LM, et al. The efficacy of bacteriophage preparations in treating inflammatory urologic diseases. Urological Nephrology. 1995;**5**:14-17

[14] Nishikawa H, Yasuda M, Uchiyama J, Rashel M, Maeda Y, Takemura I, et al. T – even related bacteriophages as candidates for treatment of *E. coli* UTI. Archives of Virology. 2008;**153**:507-515

[15] Sillankorva S, Oliveira D, Moura A, Henriques H, Faustino A, Nicolau A, et al. Efficacy of a broad host range lytic bacteriophage against *E. coli* adhered to urothelium. Current Microbiology. 2010;**62**:1128-1132

[16] Sarkar SA, McCalhin S, Barrett OC, Berger B, Pittel AC, Sultana S, et al. Oral T4 like phage cocktail application to healthy volunteers from Bangladesh. Virology. 2012;**434**:222-232

[17] Scholl D, Rogers S, Adhya S, Merril CR. Bacteriophage K1-5 encodes two different tail fiber proteins, allowing it to infect and replicate on both K1 and K5 strains of *E. coli*. Journal of Virology. 2001;**75**:2509-2515

[18] Lu TK, Collins JJ. Dispersing biofilms with engineered enzymatic bacteriophage. Proceedings of the National Academy of Science USA. 2007;**104**:11197-11202

[19] Moller-Olsen C, Ho SFS, Shukla RD, Feher T, Sagona AP. Engineered K1F bacteriophages kill intracellular *E. coli* K1 in human epithelial cells. Scientific Reports. 2018;**8**:17559

[20] Wang S, Gu J, Lv M, Guo Z, Yan G, Yu L, et al. The antibacterial activity of *E. coli* bacteriophage lysin lysep 3 is enhanced by fusing the *Bacillus amyloliquefaciens* bacteriophage endolysin binding domain D8 to the C-terminal region. Journal of Microbiology. 2017;**55**:403-408

[21] Schmelcher M, Donovan DM, Loessner MJ. Bacteriophage endolysins as novel antimicrobials. Future Microbiology. 2012;**7**:1147-1171

[22] Biers Y, Walmagh M, Van Puyenbrocck V, Comelissen A, Cemens W, Aertsen A, et al. Engineered endolysin based 'artilysins' to combat multidrug resistant Gram-negative pathogens. MBio. 2014;**5**:e01379

[23] Schirmeier E, Zimmermann P, Hofmann V, Biebl M, Gerstmans H, Maervoet VE. Inhibitory and bactericidal effect of artilysin (R) ART-175 against colistin resistant mcr-1 positive *E. coli* isolates. International Journal of Antimicrobial Agents. 2017;**51**:528-529

[24] Ma Q, Guo Z, Gao C, Zhu R, Wang S, Yu L, et al. Enhancement of the direct antimicrobial activity of Lysep 3 against *E. coli* by inserting cationic peptides into its C-terminus. Antonie Leewenhock. 2017;**110**:347-355

[25] Zampara A, Martine C, Sφrensen H, Grimson D, Antenucci F, Vitt RA, et al. Exploiting phage receptor binding proteins to enable endolysins to kill Gram-negative bacteria. Scientific Reports. 2020;**10**:12087

[26] Ryan EM, Al Kawareek MY, Donnelly RF, Gilmore BF. Synergestic phage-antibiotic combinations for the control of *E. coli* biofilms in vitro. FEMS Immunology and Medical Microbiology. 2012;**65**:395-398

[27] Verma V, Harjai K, Chhibber S. Restricting Ciprofloxacin induced resistant variant formation in biofilm of *K. pneumoniae* B5055 by complementary phage treatment. The Journal of Antimicrobial Chemotherapy. 2009;**64**:1212-1218

[28] Malik S, Nehra K, Rana JS. Bacteriophage cocktail and phage antibiotic synergism as promising alternatives to conventional antibiotic for the control of multi-drug resistant uropathogenic *E. coli*. Virus Research. 2021;**302**:198496

[29] Tanji Y, Shimada T, Yoichi M, Miyanaka K, Hori K, Unno H. Toward

rational control of *E. coli* 0157:H7 by a phage cocktail. Applied Microbiology and Biotechnology. 2004;**64**:270-274

[30] Henning U, Jann K. Two component nature of bacteriophage T4 receptor activity in *E. coli* K-12. Journal of Bacteriology. 1979;**137**:664-666

[31] Lenski RE. Two-step resistance by *E. coli* to bacteriophage T2. Genetics. 1984;**107**:1-7

[32] Doolittle MM, Cooney JJ, Caldwell DE. Lytic infection of *E. coli* biofilms by bacteriophage T4. Canadian Journal of Microbiology. 1995;**41**:12-18

[33] Pires DP, Cleto S, Sillanka VAS, Azeredo J, Lu TK. Genetically engineered phages: A review of advances over the last decade. Microbiology and Molecular Biology Reviews. 2016;**80**:523-543

[34] Loose M, Saez Moreno D, Mutti M, Hitzenhammer E, Visram Z, Dippel D, et al. Natural Bred ε^2 phages have an improved host range and virulence against UPEC over their ancestor phages. Antibiotics. 2021;**10**:1337

[35] Rodriguez-Rubio L, Gutierrez D, Donovan DM, ,Martinez B, Roriguez A, Garcia P (2016). Phage lytic proteins: Biotechnological applications beyond clinical antimicrobials. Critical Reviews in Biotechnology 36:542-552.

[36] Guo M, Feng C, Ren J, Zhuang X, Zhang Y, Zhu Y, et al. A novel antimicrobial endolysin Lys PA 26 against *P. aeruginosa*. Frontiers in Microbiology. 2017;**8**:293

[37] Oliveira H, Sao-Joe C, Azeredo J. Phage –derived peptidoglycan degrading enzymes: Challenges and future prospects for in vivo therapy. Viruses. 2018;**10**:292

[38] Chaudhary WN, Concepcion-Avecado J, Park T, Andleeb S, Bull JJ, Levin BR. Synergy and order effects of antibiotics and phages in killing *P. aeruginosa* biofilms. PLoS One. 2017;**12**:e0168615

[39] Letrado P, Corsini B, Diez-Martinez R, Bustamante N, Yuste JE, Garcia P. Bactericidal synergism between antibiotics and phage endolysin Cpe-711 to kill multidrug-resistant Pneumococcus. Future Microbiology. 2018;**13**:1215-1223

[40] Hancock V, Dahl M, Klemm P. Abolition of biofilm formation in urinary tract *E. coli* and Klebsiella isolates by metal interference through completion for Fur. Applied and Environmental Microbiology. 2010;**76**:3836-3841

Section 2

Salmonella

Chapter 3

Salmonella Infection and Pathogenesis

Kaisar Ahmad Bhat, Tasaduq Manzoor, Mashooq Ahmad Dar, Asmat Farooq, Kaisar Ahmad Allie, Shaheen Majeed Wani, Tashook Ahmad Dar and Ali Asghar Shah

Abstract

Salmonella genus represents most common food borne pathogens isolated from food producing animals and is responsible for causing zoonotic infections in humans and other animal species, including birds. As a result, *Salmonella* diseases are among the most common problems for the humans, animals, and food industry around the world. Despite rising attention about other pathogens, *Salmonella* continues to be the most prominent cause of food borne disease worldwide. *Salmonella* can be transferred to humans at any point along the farm-to-fork chain, most commonly through infected animal-derived foods such as poultry and poultry related products (eggs), pork, fish, and so on. Some *Salmonella* serotypes have been confined to a single serovar and are known as "host-restricted" while the others have a wide host spectral range and are known as "host-adapted" serotypes. Globally *Salmonella* infection causes huge mortality and the infection plays a huge role in immune response by evolving multiple mechanism to subvert immunity to its own benefit. Numerous infectivity markers and determinants have indeed been reported to play essential role in *Salmonella* pathogenesis to colonize its host by invading and avoiding the host's intestinal shielding system.

Keywords: *Salmonella*, serovars, infection, pathogenesis

1. Introduction

Salmonella is a species in the genus with worldwide public health implications and is the major cause of foodborne disease, accounting for deaths of thousands of people worldwide [1–9]. *Salmonella* is anaerobic in nature and is a Gram-negative, rod-shaped bacterium belonging to the Enterobacteriaceae family. *Salmonella* is divided into two species: *Salmonella enterica* and *Salmonella bongori*. More than 2600 *S. enterica* serovars have been defined so far, with most of these serotypes likely to cause diseases in both humans and animals [10], whereas a few *S. enterica* variants, such as *Salmonella* Gallinarum (SG) and *Salmonella* Pullorum (SP), are non-flagellated and non-motile, the large percentage of *Salmonella* members are motile by peritrichous flagella. The SG and SP are linked to clinical disease in poultry

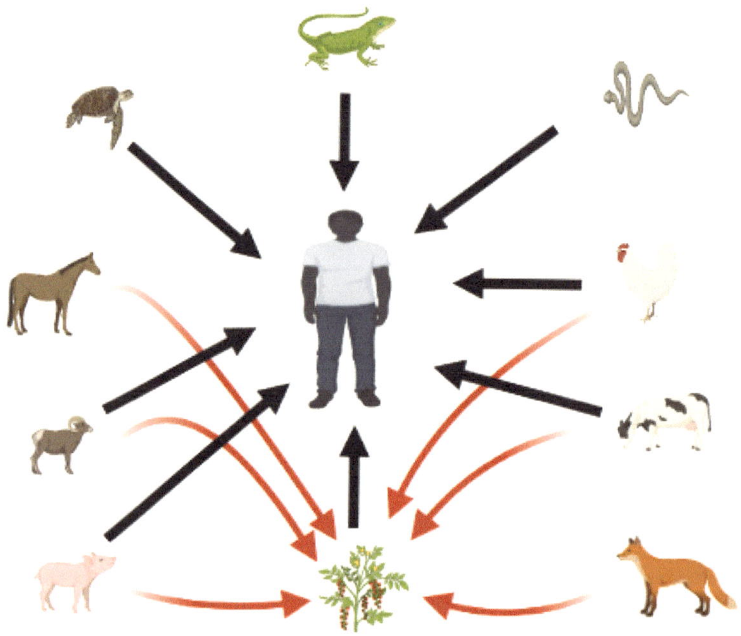

Figure 1.
Sources of Salmonella enterica.

and cause significant economic losses to poultry farming, particularly in developing countries [11–13]. According to recent data from the United States, Europe, and Low and Middle Income Countries (LMICs), *Salmonella* is frequently occurring international cause of foodborne disease. *Salmonella* also enhances food contamination in many natural environments [14]. *Salmonella enteric* found in the gut of food animals more persistently, is characterized by chronic transmitters which remove the bacterium with their own fecal matter. As a result, these carriers act as a reservoir for future bacterial contamination, allowing *Salmonella* to spread through infected milk, meat, eggs, and other agricultural products fertilized and developed in *Salmonella*-infested manure [14]. *Salmonella* have been isolated from variety of animals and their food products. These include poultry, ovine, porcine, bovine, lizards and snakes (**Figure 1**). This book chapter attempts to discuss different aspects of *Salmonella* serovars and *Salmonella* infection in different animals, with special emphasis to understand the mechanism of its pathogenesis.

2. Brief history, morphology, physical and biochemical characteristics

In mid of nineteenth century, *Salmonella* was first reported by Eberth, which was followed by Gaffky who isolated and demonstrated that Bacillus causes human typhoid fever [15]. In 1885, Theobald Smith and Daniel Elmer Salmon from the gut of pigs isolated Bacillus infected with swine fever (hog cholera) [15, 16]. An American pathologist, Dr. Daniel Elmeri Salmon, in collaboration with Smith gave the name *Salmonella* [17]. Most reference centres of *Salmonella* all over the world, including Centers for Disease Control (CDC), use *Salmonella* nomenclature system of World Health Organization (WHO) [18].

Salmonella are anaerobic, chemo-organotrophic, rod-shaped with size 0.2–1.5 × 2–5 μm and are Gram negative in nature [19]. Except a few serovars *viz S. choleraesuis*, all other members of this genus produce hydrogen sulphide and majority of them do not perform lactose fermentation [20]. This crucial trait has been used to produce a number of selective and differential media for *Salmonella* culture, isolation, and presumptive identification. *Salmonella*-Shigella agar (SS), brilliant green agar (BGA), xylose lysine deoxycholate (XLD) agar, Hektoen enteric (HE) agar, MacConkey agar, lysine iron agar (LIA), and triple sugar iron (TSI) agar are among the media frequently used [21, 22].

Salmonella is non-fastidious, as outside the living hosts it can grow and multiply in a variety of environments. *Salmonella* is heat-sensitive, and is frequently killed at temperatures of 70°C or above. The majority of serotypes thrive and grow in temperatures ranging from 5 to 47°C with an optimum of 32 to 35°C. Few serotypes, however, may thrive at temperatures as low as 2–4°C and as high as 54°C [23]. *Salmonella* grow at pH ranging from 4 to 9, with optimum range of 6.5–7.5. *Salmonella* require high water activity of about 0.99–0.94 for survival. At pH greater than 3.8, water activity greater than 0.94 and temperature higher than 70°C, it shows no growth [23]. While almost all serotypes do not make indole, hydrolyze urea, or deaminate phenylalanine or tryptophan, the majority of serotypes rapidly convert nitrate to nitrite, ferment a range of carbohydrates with acid production [20].

3. *Salmonella* nomenclature, taxonomy and serovars

The nomenclature system of Salmonella is a complex process. This genus is composed of two main species, Salmonella enterica and *Salmonella bongori*. *Salmonella enteric* is further divided into 06 subspecies on the basis of biochemical properties and genomic relatedness [24]. The subspecies are denoted by Roman numerals: I. *S. enterica* subsp. *enterica*; II. *S. enterica* subsp. *salamae*; III. *S. enterica* subsp. *arizonae*; IIIa. *S. enterica* subsp. *diarizonae*; IV. *S. enterica* subsp. *houtenae*; V. *S. enterica* subsp. *indica*. The *S. enterica* subsp. *enterica* (I) is most common subspecies of *Salmonella* and is found to be predominantly associated with around 99% of *Salmonella* infections in humans & warm blooded animals. The remaining 05 subspecies and *S. bongori* are mainly attributed to *Salmonella* infections in cold blooded animals and are rarely found in humans [25].

For serotypes in subspecies (I), CDC uses names i.e. Enteritidis, Typhimurium, Choleraesuis and Typhi while as for the unnamed serotypes described post 1966 antigenic formulae are used in subspecies II, IV, VI and *S. bongori*. The name generally refers to the location (geographic) where the serovar/serotype was isolated first. In order to avoid any confusion between species and serotype, the first letter of the named serotype is written in capital and is not italicized. At the first citation of a serotype, the genus name is given first, followed by the word “serotype” or abbreviated form “ser” and finally the serotype name is written. One of the examples is *Salmonella* serotype or ser. Typhimurium. Afterwards the genus name can be directly written followed by serotype name (e.g. *Salmonella Typhimurium* or *S. Typhimurium* [26, 27].

4. Salmonella infection

Infection with *Salmonella* causes morbidity and mortality all over the world, with the host immune response varied depending on whether the infection is acute or systemic. In addition to this, anatomical location of *Salmonella* infection plays a huge role in immune response as it evolves multiple mechanisms to subvert immunity to its own benefit.

4.1 In humans

S enterica subsp. *enterica* continues to be a leading source of disease in humans and livestock around the world. The transmission of pathogens caused a huge portion of public health and economic loss. As agricultural production began to increase after World War II, Salmonellosis become more prevalent in different countries as was the case with Europe. Despite the fact that the genus *Salmonella* contains over 2600 serovars, only 05–08 serovars cause the majority of human Salmonellosis cases in the United States. As per CDC, *Salmonella enteric* ser Enteritidis (24.7%), *S.* ser Typhimurium (23.5%), *S.* ser Newport (6.2%), and *S.* ser. Heidelberg were responsible for approximately 60% of human cases. That year, 04 serotypes accounted for 46.4% of non-human isolates. Main reason for infection in humans and other mammals is *S enteric*a which is responsible for 99% of overall infection [28]. Non-invasive non-typhoidal Salmonellosis, Invasive non-typhoidal Salmonellosis, and typhoid fever are the three principal diseases produced by *Salmonella* in humans, and these are all covered in greater depth below.

4.1.1 Non-invasive, non-typhoidal Salmonellosis

The non-typhoidal Salmonellosis (NTS) is associated with all the diseases of humans caused by *Salmonella* serotype except for the distinct typhoidal serotypes: Typhi and Paratyphi A-C. Salmonellosis is contracted orally through contaminated food or water. About 1.3 billion cases are reported annually of Salmonellosis gastroenteritis, causing huge mortality, approximately 03 million deaths globally [29]. According to the recent reports, NTS gastroenteritis is infecting developing countries. Acute enterocolitis is a symptom of Salmonellosis, and it is often followed by inflammatory diarrhea, which is only seen in people infected with invasive serovars (*S. Typhi*). The symptoms appear usually between 6 and 72 h. Primary symptoms of this disease are abdominal pain, diarrhea with or without blood, nausea, and vomiting.

4.1.2 Invasive non-typhoidal Salmonellosis

In Sub-Saharan Africa, a new *Salmonella* strain is emerging, with pathogenesis that is distinct from its genetic equivalents. This novel pathogen is known as Salmonella invasive non-typhoidal (iNTS). *Salmonella* serotypes *S. Typhimurium* and *S. Enteritidis* are the most typically connected with invasive NTS, however other serotypes such as Choleraesuis and Dublin have also been found to produce invasive illness in humans. [30, 31]. In Africa it has been found that invasive isolates have dominating genotype with several biological variations from the isolated strain (ST313) which proves that its genotype has surfaced new pathogenic clade in Sub-Saharan Africa and it may be the reason of invasive disease in humans [32]. In different parts of world other strains have also evolved which include *S. Typhimurium* ST313 strain, which gave an idea that this disease is spreading globally [33]. It was reported that iNTS are the main cause of bloodstream infections in African children [34]. Soon after the detection of AIDS in Africa, iNTS have also been reported in kids and adults and thus prompting a possible link between HIV and iNTS [34]. In New Jessey, first epidemiological link of iNTYS and AIDS was made with iNTS remained a prevalent bacterial bloodstream infection of kids and adults in Sub-Saharan Africa [32].

4.1.3 Typhoid fever

The main causative agent of Typhoid Fever is *Salmonella Typhi*. Every year about 21 million cases are being reported with almost 200,000 deaths globally. The yearly death rate increased by 39% from year 1990 to 2010 [35]. It has been reported that death rate caused by *Salmonella Typhi* in developing countries is comparatively similar to the death rate caused by breast cancer, prostate cancer, and leukemia in North America [36, 37]. Polysaccharide capsular agent allows *S. Typhi* to adapt to the acidic environment of stomach soon after infection as *S. Typhi* (acapsular) being less virulent [38, 39]. Unlike NTS, which has broad host specificity, *S. Typhi* is only found in humans. [40]. *Salmonella Typhi* inhabits and duplicates in host cells, these cells are used to translocate bacteria to liver, spleen and bone marrow. These cells include dendritic cells, neutrophils and macrophages [41].

4.2 In livestock

Salmonella infections can be seen in reptiles such as turtles, lizards, and snakes; birds like domestic pigeons and parrots; amphibians such as frogs and mammals such as dogs and cats. These infections are not frequent in small captive animals. Infection may be undetectable in reptiles, canines, and kittens although Salmonella could be identified in the stools of healthy animals. The guts of some animals can happily support these creatures which become the carrier animals of *Salmonella*. Diarrhea and enteritis are the common symptoms of Salmonellosis. Septicaemia can also be caused by *Salmonella's* invasion in the host. This intrusion causes rise in body temperature, which is usually associated with *Salmonella* infection-induced enteritis. Drowsiness, loss of appetite and diarrhea are the clinical signs of *Salmonella* infection. The diarrhea could be severe, and typically domestic dogs and cats could become extremely ill and unknowingly pollute the residence. While in birds, this disease is seldom visible. However, animals or birds that are juvenile, aged, or weak may be badly harmed by the diarrhea-induced exhaustion. They develop sepsis and expire. Most of the affected organisms may experience diarrhea for a short period of time but the majority make a full recovery. Any recuperating animal can act as a vector of infection for a period of time. *Salmonella* can dwell in low numbers in the gastrointestinal system and lymphatic system, especially in locations like caecum of birds. *Salmonella* infection may recur if the organism develops another disease [42].

4.3 In domestic fowl and poultry

Salmonella causes four types of infections in poultry, all of which are serious: Pullorum serovars of *S. enterica* causes Pullorum disease, *S. Gallinarum* causes fowl typhoid, arizonae subspecies of *S. enterica* causes arizonosis [43] and several subspecies of Salmonella like *S.* Infantis, *S.* Enteritidis and *S.* Typhimurium cause paratyphoid. The unique *S. enterica* serovars Pullorum and Gallinarum seen in poultry have been largely eliminated from European and North American industries. Nonetheless, these serovars pose a greater hazard to avian safety and wellness in areas of the globe which have low industry development, particularly in areas with inadequate protection. Despite the fact that these two serovars of *S. enterica* are normally found in chicks, spontaneous occurrences caused by these serovars. Residential chicken are among the major reservoirs of this bacteria, posing a risk to human health through the intake of contaminated foods have indeed been reported in other birds like guinea fowl and turkeys.

Poultry products have been found as a major origin of Salmonellosis on numerous occasions. Across the year 2000, an approximate 182,060 Americans were sick with *S.* Enteritidis after eating tainted eggs [44]. During 1985 and 1999, eggs were blamed for about 80,010 *S.* Enteritidis cases in the United States [45]. In addition, consumption of infected chicken has been recognized as a major potential cause for *S.* Enteritidis transmission [46]. Several of the serovars which are frequent in humans are also abundant in poultry, demonstrating the relevance of livestock as a source for the spread of *Salmonella* in people [47]. *Salmonella's* potential to infect chicken is highly linked to the transmitting serotype, as well as the maturity and genetic lineage of the bird. The disease caused by Gallinarum serovar of *S. enterica*, Fowl Typhoid (FT) spreads mostly through fecal-oral route [48]. There are also diseases mostly restricted to the gut caused by different *Salmonella* serovars in poultry [43]. Salmonellosis is the most common symptom of Typhimurium serovar infection in small birds. Fatality rates differ greatly, ranging from as low as 10% to as high as 80% in extreme cases.

4.4 In cattle

Salmonellosis is a leading cause of death and disease in livestock, some of which are commonly detected which are infected sub-clinically. As a result, cattle serve as a significant storehouse for diseases infecting humans. Several studies have been published during last decade with an emphasis on multi drug resistance variants and significance of *Salmonella* for food sector [49, 50]. Surprisingly, although extensive research was done on Salmonellosis, the infection and its associated risks remain unaddressed [43]. Salmonellosis is still a disease that affects livestock all over the world and is largely caused by the *S.* Dublin and *S.* Typhimurium. Additional serotypes have been linked to cattle infections on a stochastic basis [48]. Studies documented the identification of 101 distinct serotypes of Salmonella in cattle, most of which had a reduced incidence [43, 51]. In late 1960s, *Salmonella* infections in the livestock sector of Britain peaked with over 4000 cases reported in 1969 [48, 51]. Seven (07) serovars of *Salmonella* were found in 48% of the 730 isolated *Salmonella* from livestock in the United States [50]. There is a risk of novel strains being imported which was reported in United Kingdom as 10 *Salmonella* serovars were identified which were of non GB origin [43].

4.5 In pigs and sheep

S. enterica serovars Choleraesuis was first detected in swine when it was thought to be the causative agent of swine fever (hog cholera). The susceptibility of swine to *Salmonella* is determined by a number of parameters like the infecting serotypes and the pig's age. Further, the incidence of Salmonellosis varies from region to region and is weakly linked to swine population, farming techniques, and their mixing [43]. *Salmonella* serovars linked to clinical illness in swine can be separated into two categories: Choleraesuis like host specific serovars and *S.* Typhimurium like ubiquitous. However, the presence of *S.* Choleraesuis has substantially decreased since then, and it is currently only spotted occasionally whereas Typhimurium still remains a severe threat to the swine sector especially in United States. Several serovars like Typhimurium, Copenhagen, Agona, Derby and Heidelberg were by far the most prevalent serovars in swine in the United States in the first decade of this century. Three of these serovars were isolated from humans during this time span [52]. During last 02 decades, research studies on other serovars have increased either due to improved surveillance or due to increased occurrence of infection.

Sheep Salmonellosis appears to be frequent in nations with considerable sheep population, including United Kingdom, Australia, New Zealand, and the United States of America. The seasonality of Salmonellosis spread and the incidence of diseases caused by widespread serovars is usually linked to sheep mobility and transportation [53, 54]. The Ovis strains of Serovar Abortus with restricted hosts are predicted to be introduced into diseased sheep flocks and spread *via* the fecal oral route [55]. However, there is no strong evidence that bacteria are transferred by drinking, nutrition, or the wastes of other hosts. Transmission of grazing livestock through the nasal channel may be possible due to many serovars causing pneumonia in lamb. Pulmonary discharge may transmit the bacteria to other animals.

4.6 In horses, dogs and cats

Salmonella Typhimurium was initially reported as the causative agent of Colitis in late 1910 and subsequently prevailed as a cause of Salmonellosis in horses throughout the world. Antibiotic use in conjunction with hospitalization stresses has been shown to have a significant impact on the horse's sensitivity to *Salmonella* infection. The only host suitable for hooved animals is *Salmonella* Abortusequi, which causes horse paratyphoid disease. The surge and decline in prevalence of disease by distinct serovars has become a significant characteristic of the epidemiology of horse salmonellosis in the United States. This could lead to an increase in herd immunity and decrease in the pathogenicity of the individual serovar.

Salmonella infection in cats and dogs can be subclinical, with just occasional shedding. The infection fluctuates, ranging from moderate to severe gastritis, with the possibility of miscarriage, systemic dissemination, or sepsis [56]. *Salmonella* can be excreted over a month by healed animals, and persistent transmission with intervals of re-emergence is conceivable. *Salmonella* have the ability to propagate zoonotic infections and may play a role in the establishment of antibiotic resistance in bacteria [57]. The majority of the infestations were medically quiet, however some developed moderate diarrhea. Recent research has shown that dogs fed with uncooked meat can eliminate the bacterium in their stools for a longer period.

5. *Salmonella* pathogenesis

The favorable outcome of a pathogen is based on its capability to enter a host, evade host defense barrier and initiate infection. *Salmonella* has developed contrasting schedule to destabilize normal host cellular functions that allow it to get involved in and multiple inside the host cell. Depending upon the serotype of *Salmonella* involved and health status of human host, the acuteness of *Salmonella* infection varies. Elderly people, immune-suppression patients and children below 05 years of age are more prone to *Salmonella* infection. The ability of *Salmonella* to invade, replicate and remain alive within the human host makes it more morbific that finally results into harmful mortal disease.

Salmonella produces different virulence factors that play an important role in its pathogenicity. These involve (1) the potential to invade the cell (2) a perfect lipopolysaccharide coat (3) to replicate intra-cellularly and (4) feasibly the secretion of toxins [58]. The organisms establish a colony in ileum and colon after ingestion followed by occupying the intestinal epithelium and grows rapidly within the epithelium and lymphoid follicles. *Salmonella* invasion mechanism is partially understood. On epithelial cell surface there is the presence of specific receptors. When the organism incursion occurs, enterocyte membrane goes through disarrangement that results in pinocytosis of organism. Invasion depends on rearrangement of cell cytoskeleton and may be entailed to increase in cellular inositol phosphate and calcium. After invasion, organism has ability to proliferate intra-cellularly thereby escalating to mesenteric lymph nodes and all over the body by systematic circulation; absorbed by reticulo-endothelial cells that limits and checks the expansion of an organism. There is a perceptible genetic control involving multiple genes in both chromosomes and plasmids for attachment and invasion. Some organisms has the ability to infect liver, spleen, gall bladder, bone, meninges etc. depending upon the host defense. Human *Salmonella* (gastroenteritis) resides in intestine. However, most serotypes get perished on time. After invading the intestine, most of *Salmonellae* brings on an acute inflammatory response that may lead to ulceration, also they might elaborate cytotoxins that forbid protein synthesis. It is not clear if these cytotoxins play a role in the inflammatory response or ulceration. On the other hand, invasion of the mucosa induces epithelial cells to produce and release pro-inflammatory cytokines such as IL-1, IL-6, IL-8, TNF-2, IFN-U, MCP-1, and GM-CSF. These trigger an acute inflammatory response in the body and may also be accountable to harm the intestine [59]. Due to the inflammatory reaction, symptoms such as fever, chills, stomach pain, leukocytosis, and diarrhea are frequent. Polymorphonuclear leukocytes, blood, and mucus may be seen in the stool.

One of the features of *Salmonella* is non-phagocytic nature on human host cells during invasion [60], where it literally induces its own phagocytosis in order to gain access to its host cell. *Salmonella* pathogenicity islands (SPIs), gene clusters positioned at the major chromosomal DNA region and encoding for the structures required in the invasion activity, provide the remarkable genetics that enable this brilliant technique [61]. Bacteria tend to infiltrate the epithelial cells of the intestinal wall when they enter the digestive tract *via* contaminated water or food. Type III secretion systems, or SPIs, are multi-channel proteins that allow *Salmonella* to infuse its effectors into the cytoplasm *via* the intestinal epithelial cell membrane. The bacterial effectors subsequently activate the signal transduction pathway and lead the host cell's actin cytoskeleton to be rebuilt, causing the epithelial cell membrane to ruffle outward and engulf the bacteria. The membrane ruffle's morphology is similar to the process of phagocytosis [62].

The ability of the *Salmonella* strains to remain in the host cell is important for pathogens as strains lacking this capability are non-virulent [63]. After the host cell engulfs *Salmonella*, the bacterium is enclosed in a membrane compartment called a vacuole, which is formed of the host cell membrane. The presence of the bacterial foreign body activates the host cell immune response under normal circumstances, which result in the fusion of the lysosomes and the secretion of the digestive enzymes to break down the intracellular bacteria. Although, *Salmonella* uses the type III secretion system to inject other effector proteins into the vacuole, it causes the modification of the compartment structure. The re-assembled vacuole obstructs the fusion of the lysosomes and this allows the intracellular survival and replication of the bacteria inside the host cells. The ability of the bacteria to continue within macrophages allows them to be carried in the reticulo-endothelial system (RES) [64].

The mechanisms of *Salmonella* gastroenteritis and diarrhea are well known now. Only strains that infiltrate the intestinal mucosa are associated with the appearance of an acute inflammatory reaction and diarrhea; the secretion of fluid and electrolytes by the small and large intestines causes the diarrhea. Even though, the secretion is not just an indication of tissue destruction and ulceration, the mechanisms of secretion are indistinct. Unlike *Shigella* and invasive *Escherichia coli*, *Salmonella* infiltrates the intestinal epithelial cells but, do not escape the phagosome. Therefore, the extent of intercellular spread and ulceration of the epithelium is much less. From the basal side of epithelial cells, *Salmonella* escapes into the lamina propria. Systemic spread of the organisms can occur that causes the enteric fever. Following the invasion of the intestinal mucosa, activation of mucosal adenylate cyclase occurs; that results in the increase in cyclic AMP that causes secretion. It is not understood that by which mechanism adenylate cyclase is stimulated; it might involve local production of prostaglandins or other components of the inflammatory reaction.

6. Conclusion

Globally, Salmonellosis is the main cause of bacterial disease in all living creatures. All over the world it is posing very serious public health concerns and compromising the yield and output of animal husbandry production. The effort of isolating, identifying, and reporting *Salmonella* serotypes must continue for diagnostic, therapeutic, and public health objectives, despite the fact that the nomenclature for *Salmonella* is constantly evolving and the argument over the naming for the type species is still ongoing. *Salmonella* outbreaks have been linked to a variety of foods, and researchers are scrambling to figure out how this infection impacts humans and animals. This infection is a leading source of morbidity and mortality worldwide, with the host immune response differing depending on nature of infection. The genetic makeup of *Salmonella* made it possible for its strains to adapt to different environmental conditions. The implications of this infectious disease in humans vary depending on its serotype and the health level of the human host. Thus for better understanding the genetics of *Salmonella* and to investigate the mechanisms that contribute to pathogenesis evolution, a lot of work has been done. Occurrence of two different, potentially complementary evolutionary approaches to host range and virulence were evaluated using genome sequencing of *Salmonella* serovars. It includes horizontal gene transfer, gene loss which actually affects its ability to colonize. Gene acquisition by horizontal transfer (associated with SPIs, transposable elements, phages, and plasmids) and gene loss or loss of function, which affects host range. In spite of the presence a greater amount of research findings related to *Salmonella* infection and pathogenesis mechanism in host animals, several key queries remain intact *viz* the exact role of virulence genes and genomic islands of particular serovar in animal models. Thus the need of hour is to have an in depth understanding of *Salmonella* pathogenesis for developing intervention strategy to minimize the disease's prevalence and spread, as well as assisting in the production of novel drugs and treatments which might lead to improved treatment of Salmonellosis in living creatures.

Conflict of interest

The authors declare that they have no competing interests.

Author details

Kaisar Ahmad Bhat[1,2], Tasaduq Manzoor[3], Mashooq Ahmad Dar[3,4*], Asmat Farooq[2], Kaisar Ahmad Allie[1], Shaheen Majeed Wani[1], Tashook Ahmad Dar[1] and Ali Asghar Shah[1]

1 School of Biosciences and Biotechnology, Baba Ghulam Shah Badshah University, Rajouri, Jammu and Kashmir, India

2 Proteomics Laboratory, Division of Plant Biotechnology, Sher-e-Kashmir University of Agricultural Sciences and Technology—Kashmir, India

3 Faculty of Veterinary Sciences and Animal Husbandry, Division of Biotechnology, Sher-e-Kashmir University of Agricultural Sciences and Technology—Kashmir, India

4 Department of Biochemistry/Clinical Biochemistry, University of Kashmir, Kashmir, India

*Address all correspondence to: darmashooq1@gmail.com

References

[1] Lee KM, Runyon M, Herrmanm TJ, Phillips R, Hsieh J. Review of Salmonella detection and identification methods: Aspects of rapid emergency response and food safety. Food Control. 2015; **47**:264-276

[2] Newell DG, Koopmans M, Verhoef L, Duizer E, Aidara-Kane A, Sprong H, et al. Food-borne diseases the challenges of 20 years ago still persist while new ones continue to emerge. International Journal of Food Microbiology. 2010; **139**(1):S3-S15

[3] Putturu R, Eevuri T, Ch B, Nelapati K. Salmonella enteritidis-foodborne pathogen—A review. International Journal of Pharma and Bio Sciences. 2015;**5**(1):86-95

[4] Alzwghaibi AB, Yahyaraeyat R, Fasaei BN, Langeroudi AG, Salehi TZ. Rapid molecular identification and differentiation of common Salmonella serovars isolated from poultry, domestic animals and foodstuff using multiplex PCR assay. Archives of Microbiology. 2018;**200**(7):1009-1016

[5] Eguale T. Non-typhoidal Salmonella serovars in poultry farms in central Ethiopia: Prevalence and antimicrobial resistance. BMC Veterinary Research. 2018;**14**(1):217

[6] Zishiri OT, Mkhize N, Mukaratirwa S. Prevalence of virulence and antimicrobial resistance genes in Salmonella spp. isolated from commercial chickens and human clinical isolates from South Africa and Brazil. The Onderstepoort Journal of Veterinary Research. 2016;**83**(1):e1-e11

[7] Velasquez CG, MacKlin KS, Kumar S, Bailey M, Ebner PE, Oliver HF, et al. Prevalence and antimicrobial resistance patterns of Salmonella isolated from poultry farms in Southeastern United States. Poultry Science. 2018;**97**(6):2144-2152

[8] Ed-dra A, Filali FR, Karraouan B, El Allaoui A, Aboulkacem A, Bouchrif B. Prevalence, molecular and antimicrobial resistance of Salmonella isolated from sausages in Meknes, Morocco. Microbial Pathogenesis. 2017; **105**:340-345

[9] Odoch T, Wasteson Y, L'Abée-Lund T, Muwonge A, Kankya C, Nyakarahuka L, et al. Prevalence, antimicrobial susceptibility and risk factors associated with non-typhoidal Salmonella on Ugandan layer hen farms. BMC Veterinary Research. 2017;**13**(1):365

[10] Mezal EH, Sabol A, Khan MA, Ali N, Stefanova R, Khan AA. Isolation and molecular characterization of Salmonella enterica serovar Enteritidis from poultry house and clinical samples during 2010. Food Microbiology. 2014;**38**:67-74

[11] Xiong D, Song L, Pan Z, Jiao X. Identification and discrimination of Salmonella enterica serovar Gallinarum biovars Pullorum and Gallinarum based on a one-step multiplex PCR assay. Frontiers in Microbiology. 2018;**9**:1718

[12] PenhaFilho RAC, Ferreira JC, Kanashiro AMI, Darini ALC, Berchieri JA. Antimicrobial susceptibility of Salmonella gallinarum and Salmonella Pullorum isolated from ill poultry in Brazil. Ciência Rural. 2016;**46**(3):513-518

[13] Eriksson H, Söderlund R, Ernholm L, Melin L, Jansson DS. Diagnostics, epidemiological observations and genomic subtyping

in an outbreak of pullorum disease in non-commercial chickens. Veterinary Microbiology. 2018;**217**:47-52

[14] Yue M, Schifferli DM. Allelic variation in Salmonella: An underappreciated driver of adaptation and virulence. Frontiers in Microbiology. 2013;**4**:419

[15] Le Minor L. The genus Salmonella. In: Balows A, Truper HG, Dworkin C, Harder W, Scheilfer KH, editors. The Prokaryotes. New York: Springer-Varlag; 1991. pp. 2760-2774

[16] D'Aoust JY. In: Doyle MP, editor. Salmonella. Foodborne Bacterial Pathogens. New York: Marcel Inc; 1989. pp. 327-445

[17] Eng SK, Pusparajah P, AbMutalib NS, Ser HL, Chan KG, Lee LH. Salmonella: A review on pathogenesis, epidemiology and antibiotic resistance. Frontiers in Life Science. 2015;**8**(3):284-293

[18] Popoff MY, Bockemühl J, Gheesling LL. Supplement 2001 (no. 45) to the Kauffmann-White scheme. Research in Microbiology. 2003;**154**(3): 173-174

[19] Popoff MY, Le Minor L. In: Brenner DJ, Kreig NR, Staley JT, editors. Salmonella. Bergey's Manual of Systematic Bacteriology. 2nd ed. New York, USA: Springer; 2005. pp. 764-799

[20] Rambach A. New plate medium for facilitated differentiation of Salmonella spp. From Proteus spp, other enteric bacteria. Applied and Environmental Microbiology. 1990;**56**(1):301-303

[21] Andrews WH, Jacobson A, Hammack T. BAM Chapter 5: Salmonella. Bacteriological Analytical Manual. 2007;**21**:5-01

[22] Anderson RC, Ziprin RL. Bacteriology of salmonella. Foodborne Disease Handbook. 2001;**1**:247-263

[23] Pui CF, Wong WC, Chai LC, Tunung R, Jeyaletchumi P, Hidayah N, et al. A foodborne pathogen. International Food Research Journal. 2011;**18**(2):1

[24] Reeves MW, Evins GM, Heiba AA, Plikaytis BD, Farmer JJ 3rd. Clonal nature of Salmonella typhi and its genetic relatedness to other salmonellae as shown by multilocus enzyme electrophoresis, and proposal of Salmonella bongori comb. nov. Journal of Clinical Microbiology. 1989;**27**(2):313-320

[25] Brenner FW, Villar RG, Angulo FJ, Tauxe R, Swaminathan B. Salmonella nomenclature. Journal of Clinical Microbiology. 2000;**38**(7):2465-2467. DOI: 10.1128/JCM.38.7.2465-2467.2000

[26] Popoff MY, Bockemühl J, Brenner FW. Supplement 1998 (no. 42) to the Kauffmann-White scheme. Research in Microbiology. 2000;**151**(1): 63-65

[27] Popoff MY, Le Minor L. Antigenic formulas of the Salmonella serovars. Annales de Microbiologie. 1982;**133B**:223-243

[28] Mascellino MT, editor. Salmonella: A Re-emerging Pathogen. IntechOpen, Sapienza University; 2018. DOI: 10.5772/intechopen.69799

[29] Kurtz JR, Goggins JA, McLachlan JB. Salmonella infection: Interplay between the bacteria and host immune system. Immunology Letters. 2017;**190**:42-50. DOI: 10.1016/j.imlet.2017.07.006

[30] Jean SS, Wang JY, Hsueh PR. Bacteremia caused by Salmonella enterica serotype Choleraesuis in Taiwan. Journal of Microbiology, Immunology, and

Infection.Weimianyugan ran zazhi. 2006;**39**(5):358-365

[31] Fang FC, Fierer J. Human infection with Salmonella dublin. Medicine. 1991;**70**(3):198-207

[32] Feasey NA, Dougan G, Kingsley RA, Heyderman RS, Gordon MA. Invasive non-typhoidal salmonella disease: An emerging and neglected tropical disease in Africa. The Lancet. 2012;**379**(9835):2489-2499. DOI: 1016/S0140-6736(11)61752-2

[33] Almeida F, Seribelli AA, da Silva P, Medeiros MI, dos Prazeres RD, Moreira CG, et al. Multilocus sequence typing of Salmonella Typhimurium reveals the presence of the highly invasive ST313 in Brazil. Infection, Genetics and Evolution. 2017;**51**:41-44. DOI: 10.1016/j.meegid.2017.03.009

[34] Duggan MB, Beyer LA. Enteric fever in young Yoruba children. Archives of Disease in Childhood. 1975;**50**(1):67-71. DOI: 10.1136/adc.50.1.67

[35] Crump JA, Luby SP, Mintz ED. The global burden of typhoid fever. Bulletin of the World Health Organization. 2004;**82**:346-353

[36] Gunn JS, Marshall JM, Baker S, Dongol S, Charles RC, Ryan ET. Salmonella chronic carriage: Epidemiology, diagnosis, and gallbladder persistence. Trends in Microbiology. 2014;**22**(11):648-655. DOI: 10.1016/j.tim.2014.06.007

[37] Lozano R, Naghavi M, Foreman K, Lim S, Shibuya K, Aboyans V, et al. Global and regional mortality from 235 causes of death for 20 age groups in 1990 and 2010: A systematic analysis for the Global Burden of Disease Study 2010. The Lancet. 2012;**380**(9859):2095-2128. DOI: 10.1016/S0140-6736(12)61728-0

[38] Parry CM, 1Tr H, Douqan G, et al. Typhoid fever. New England Journal. 2002;**347**:1770-1782. DOI: 10.1056/NEJMra020201

[39] Waddington CS, Darton TC, Pollard AJ. The challenge of enteric fever. The Journal of Infection. 2014;**68**:S38-S50. DOI: 10.1016/j.jinf.2013.09.013

[40] Edsall G, Gaines S, Landy M, Tigertt WD, Sprinz H, Trapani RJ, et al. Studies on infection and immunity in experimental typhoid fever I. Typhoid fever in chimpnzees orally infected with Salmonella typhosa. The Journal of Experimental Medicine. 1960;**112**(1):143-166. DOI: 10.1084/jem.112.1.143

[41] Rowe B, Ward LR, Threlfall EJ. Multidrug-resistant Salmonella typhi: A worldwide epidemic. Clinical Infectious Diseases. 1997;**24**(Supplement_1):S106-S109

[42] Salmonellosis [Internet]. 2008. Available from: http://www.pethealthcouncil.co.uk

[43] Barrow PA, Methner U. Salmonella in Domestic Animals. 2nd ed. Germany: CABI; 2013. DOI: 10.1079/9781845939021.0000

[44] Schroeder CM, Naugle AL, Schlosser WD, Hogue AT, Angulo F, Rose S, et al. Estimate of illnesses from Salmonella Enteritidis in eggs, United States, 2000. Emerging Infectious Diseases. 2005;**11**:113-115. DOI: 10.3201/eid1101.040401

[45] Patrick ME, Adcock PM, Gomez TM, Altekruse SF, Holland BH, Tauxe RV, et al. Salmonella Enteritidis infections, United States, 1985-1999. Emerging Infectious Diseases. 2004;**10**:1-7. DOI: 10.3201/eid1001.020572

[46] Kimura AC, Reddy V, Marcus R, Cieslak PR, Mohle-Boetani JC, Kassenborg HD, et al. Chicken consumption is a newly identified risk factor for sporadic Salmonella enterica serotype Enteritidis infections in the United States: A case-control study in FoodNet sites. Clinical Infectious Diseases. 2004;**38:244**-252. DOI: 10.1086/381576

[47] Gast RK. Salmonella infections. In: Saif YM, editor. Diseases of Poultry. 12th ed. USA: Blackwell Publishing; 2008. pp. 619-674. DOI: 10.7589/0090-3558-45.1.251

[48] Mastroeni P, Maskell D. Salmonella Infections, Clinical, Immunological and Molecular Aspects. 1st ed. UK: Cambridge University Press; 2006. DOI: 10.1017/CBO9780511525360.002

[49] Mohler VL, Izzo MM, House JK. Salmonella in calves. Veterinary Clinics: Food Animal Practice. 2009;**25**:37-54. DOI: 10.1016/j.cvfa.2008.10.009

[50] Alexander KA, Warnick LD, Wiedmann M. Antimicrobial resistant Salmonella in dairy cattle in the United States. Veterinary Research Communications. 2009;**33**:191-209. DOI: 10.1007/s11259-008-9170-7

[51] Sojka WJ, Wray C, Shreeve J, Benson JA. Incidence of Salmonella infections in animals in England and Wales, 1968-74.The Journal of Hygiene 1977;78:43-56. DOI: 10.1017/S0022172400055923

[52] Haley CA, Dargatz DA, Bush EJ, Erdman MM, Fedorka-Cray PJ. Salmonella prevalence and antimicrobial susceptibility from the National Animal Health Monitoring System Swine 2000 and 2006 studies. Journal of Food Protection. 2012;**75**:428-436. DOI: 10.4315/0362-028X.JFP-11-363

[53] Richards RB, Norris RT, Dunlop RH, McQuade NC. Causes of death in sheep exported live by sea. Australian Veterinary Journal. 1989;**66**:33-38. DOI: 10.1111/j.1751-0813.1989. tb03011.x

[54] Higgs AR, Norris RT, Richards RB. Epidemiology of salmonellosis in the live sheep export industry. Australian Veterinary Journal. 1993;**70**:330-335. DOI: 10.1111/j.1751-0813.1993.tb00874.x

[55] Jack EJ. Salmonella abortusovis: An atypical Salmonella. The Veterinary Record. 1968;**82**:558-561

[56] Stiver SL, Frazier KS, Mauel M, Styer EL. Septicemic salmonellosis in two cats fed a raw meat diet. Journal of the American Animal Hospital Association. 2003;**39**:538-542. DOI: 10.5326/0390538

[57] Lynne AM, Dorsey LL, David DE, Foley SL. Characterization of antibiotic resistance in host adapted Salmonella enterica. International Journal of Antimicrobial Agents. 2009;**34**:169-172. DOI: 10.1016/j.ijantimicag.2009.02.018

[58] Giannella RA. Salmonella. In: Baron S, editor. Medical Microbiology. 4th ed. Galveston (TX): University of Texas Medical Branch at Galveston; 1996. p. 21

[59] Peterson JW. Bacterial pathogenesis. In: Baron S, editor. Medical Microbiology. 4th ed. Galveston (TX): University of Texas Medical Branch at Galveston; 1996. p. 7

[60] Hansen-Wester I, Stecher B, Hensel M. Type III secretion of *Salmonella enteric serovars* Typhimuriumtranslocated effectors and SseFG. Infection and Immunity. 2002;**70**:1403-1409. DOI: 10.1128/IAI.70.3.1403-1409.2002

[61] Grassl GA, Finlay BB. Pathogenesis of enteric *Salmonella* infections.

Current Opinion in Gastroenterology. 2008;**24**:22-26. DOI: 10.1097/MOG.0b013e3282f21388

[62] Takaya A, Suzuki M, Matsui H, Tomoyasu T, Sashinami H, Nakane A, et al. Stress-induced ATP-dependent protease, is critically important for systemic *Salmonella enterica* serovar typhimurium infection of mice. Infection and Immunity. 2003;**71**:690-696. DOI: 10.1128/IAI.71.2.690-696.2003

[63] Bakowski MA, Braun V, Brumell JH. Salmonella-containing vacuoles: Directing traffic and nesting to grow. Traffic. 2008;**9**:2022-2031. DOI: 10.1111/j.1600-0854.2008.00827.x

[64] Monack DM, Mueller A, Falkow S. Persistent bacterial infections: The interface of the pathogen and the host immune system. Nature Reviews. Microbiology. 2004;**2**:747-765. DOI: 10.1038/nrmicro955

Chapter 4

Salmonella in Wild Animals: A Public Health Concern

Eliege Jullia Eudoxia dos Santos,
Amanda Teixeira Sampaio Lopes and Bianca Mendes Maciel

Abstract

Wildlife can be a reservoir of infectious agents for humans and domestic and wild animals. In this regard, widespread *Salmonella* spp. in wildlife is a problem for public and environmental health. Currently, more than 2500 serovars of *Salmonella* spp. are widely distributed among humans, animals, and the environment. This ubiquity favors the bidirectional transmission of the pathogen between wild and domestic animals. Moreover, when farmed animals acquire *Salmonella* spp. from wildlife, the likelihood of humans becoming infected increases. The risk is higher in forest environments impacted by human activities or when animals are removed from their natural habitat. Consequently, human contact with wild animals in captivity increases the risk of salmonellosis outbreaks. These animals are often carriers of *Salmonella* spp. strains multiresistant to antibiotics, which makes it difficult to treat and control the disease. Therefore, prevention and control measures of this pathogen must include both the pathogen-host relationship and the environment, with a surveillance system for emerging and re-emerging diseases from wildlife.

Keywords: wild fauna, *Salmonella* asymptomatic carrier, salmonellosis, anthropized forest environment, wild animals in captivity

1. Introduction

Natural environments have been altered by the destruction of forests and habitats to expand habitable zones for humans. These changes expose humans and animals to infectious agents that were restricted to certain species and geographical areas. Furthermore, these changes cause an epidemiological, sanitary, and environmental rearrangement of diseases, especially those with zoonotic profiles, as in the case of salmonellosis [1].

Salmonella spp. is a bacterium with pathogenic characteristics often associated with food infections and outbreaks, with serious public health implications. Infections can affect people, livestock such as cattle, pigs, sheep, poultry, pets, and even wild animals. Concerning wild animals, the characteristic ubiquity of the bacterium also favors cross-contamination to domestic animals, especially in areas for livestock close to forests.

Epidemiologically, one of the main characteristics of *Salmonella* is its condition as a latent carrier [2]. Latency corresponds to a state in which the individual does not present clinical symptomatology, but continues eliminating the agent intermittently in the feces. Thus, these asymptomatic latent carriers become natural reservoirs and, consequently, maintainers of the pathogen both in the food chain and in the environment.

Naturally, wild animals can be asymptomatic carriers of *Salmonella* spp., with the bacterium remaining in equilibrium with the intestinal microbiota. When these animals are kept away from their natural habitat, the resulting stress compromises their immune system and destabilizes the microbiota, leading to increased elimination of the pathogen in feces. Therefore, wild animals kept in captivity tend to have a higher prevalence of *Salmonella* spp. than free-living animals, possibly leading to outbreaks of salmonellosis in humans due to cross-contamination by serotypes of *Salmonella* spp. This scenario is even worse when the serotype involved is multidrug-resistant to antibiotics.

The maintenance of wild animals in captivity is a major public health concern, especially in the case of reptiles. We conducted a study with fecal samples of 30 tegu lizards born in captivity that were asymptomatic latent carriers of *Salmonella* spp., with nine serotypes with resistance to at least two antibiotics being isolated [3]. In another study using 31 snakes kept in captivity, 58% tested positive for *Salmonella* spp. and seven serotypes were isolated [4]. Some of the animals, both among the tegus and the snakes, tested positive for more than one serotype with different resistance profiles. In preserved forest areas, the prevalence of *Salmonella* spp. in wild animals is usually lower. Our research team sampled 518 free-living wild animals in forest fragments (388 mammals, 114 birds, and 16 reptiles) from 2015 to 2021 in four mesoregions of Bahia (north-central Bahia, south-central Bahia, Metropolitan Salvador, and south Bahia), Brazil, and observed that only three mammals (unpublished data) and one bird [5] tested positive for *Salmonella* spp.

Notably, the manifestation of salmonellosis is associated with factors inherent to the etiological agent, the host, and the environment. The correlation between the three will determine the impacts on biosecurity and persistence of the bacterium in ecosystems, food, and carriers. The prevention and control of this pathogen demand interdisciplinary and international cooperation based on shared data to ensure a more effective approach to outbreaks.

2. General characteristics of the genus *Salmonella*

Salmonella is a genus of pathogenic bacteria named by Lignières in 1900, after the veterinarian pathologist and microbiologist Daniel Elmer Salmon, who isolated the agent and associated it with a disease for the first time [6]. These bacteria are part of the Enterobacteriaceae family and are morphologically composed of non-spore-forming, Gram-negative, facultatively anaerobic, rod-shaped bacteria, with optimum growth temperature between 35°C and 37°C [7].

Currently, *Salmonella* spp. is divided into two species: *Salmonella enterica* and *Salmonella bongori*. The first species is divided into six subspecies with a Roman numeral, as follows: *enterica* (serogroup I), *salamae* (serogroup II), *arizonae* (serogroup IIIa), *diarizonae* (serogroup IIIb), *houtenae* (serogroup IV), and *indica* (serogroup VI) [8, 9].

Salmonella bongori (serogroup V) has 23 serotypes and *S. enterica* has more than 2500 serotypes (*S. enterica* subsp. *enterica* serogroup I = 1547, *S. enterica* subsp. *salamae* serogroup II = 513, *S. enterica* subsp. *arizonae* serogroup IIIA = 100, *S. enterica* subsp. *diarizonae* serogroup IIIb = 341, *S. enterica* subsp. *houtenae* serogroup IV = 73, and *S. enterica* subsp. *indica* serogroup VI = 13 [10]. This characterization of species and subspecies into serotypes is based on the model proposed by Kauffman-White from differences observed in flagellar (H), capsular (K), and somatic (O) antigens [11].

Species	*Salmonella enterica*						*Salmonella bongori*
Subspecies	*enterica*	*salamae*	*arizonae*	*diarizonae*	*houtenae*	*indica*	
Dulcitol	+	+	—	—	—	*	+
Malonate	—	+	+	+	—	—	—
Gelatinase	—	+	+	+	+	—	+
Sorbitol	+	+	+	+	+	—	+
Galacturonate	—	+	—	+	+	+	+
Salicin	—	—	—	—	+	—	—

**Variable according to serovar.*

Table 1.
Biochemical characteristics of Salmonella *species and subspecies.*

The species and subspecies of *Salmonella* also have distinguishing biochemical characteristics (**Table 1**). These bacteria are catalase-positive and oxidase-negative and can form hydrogen sulfide through the enzyme cysteine desulfhydrase, which promotes sulfur reduction. Moreover, they can reduce nitrite to nitrate and use citrate as an energy source. In contrast, they do not produce indole or hydrolyze urea [6].

Salmonella is a bacterium of worldwide geographical distribution and, therefore, many animal species, including wild animals, can act as a reservoir of its various serovars [12]. Wild and domestic animals and humans can be affected by any of the more than 2500 different serovars [13]. *S. enterica* subsp. *enterica* determines infections mainly in warm-blooded animals [11], chiefly mammals [14], and is associated with most of the world's foodborne diseases [11]. Nevertheless, different serovars of this subspecies have been isolated from exotic reptile kept as pets, as we will report throughout this chapter. The other subspecies of *S. enterica* are uncommon for humans and are usually found in cold-blooded animals and environmental samples [14]. Similarly, *S. bongori* is more common in cold-blooded animals, especially reptiles, and in the environment [6], but can also infect humans [15].

Salmonella habitat, based on the host's specificity and clinical manifestations, can be characterized as follows: a. *highly adapted to humans*, corresponding to serotypes *S.* Typhi, and *S.* Paratyphi A, B, and C; b. *highly adapted to animals*, responsible for paratyphoid fever in animals, consisting of *S.* Dublin (cattle), *S.* Choleraesuis and *S.* Typhisuis (pigs), *S.* Abortusequi (equines), and *S.* Pullorum and *S.* Gallinarum (birds); and c. *zoonotic Salmonella*, which affect humans and domestic and wild animals indistinctly and are involved in food poisoning and gastroenteritis. This third group is more representative of public health due to its high morbidity and mortality rates [6, 9].

Notably, *Salmonella* spp. can survive in the environment, mainly in organic matter, and can continue infecting for 280 days in soils used for cultivation, 120 days in pastures, 30 days in bovine feces, and 28 days in bird fecal matter [6, 16]. Moreover, it adheres to the surface of plant roots and survives for long period underground [17]. This occurs because these bacteria, which inhabit the intestinal tract of humans and animals, are eliminated in the feces and can then contaminate both water and soil. Furthermore, in aquatic ecosystems, *Salmonella* can adhere to sediments [18] and survive in high densities in these systems and water after 56 days [19]. In this regard, sediments provide a protective layer for enteric bacteria from a nutrient reserve and prevent stress from the aquatic environment [20].

3. *Salmonella* spp. in wild animals

The increased demand for wild animals raised as domestic animals has become a public health and environmental concern due to the spreading of pathogens [21].

Class	Year	Animal	Pet (P) Free (F)	*Salmonella* serovar	Cases: numbers of Illnesses (I); Hospitalizations (H); Deaths (D)			Ref.
					I	H	D	
Reptiles	2021	Turtle	P	*S.* Typhimurium	87	32	01	[22]
	2020	Bearded Dragon	P	*S.* Muenster	18	11	00	[22]
	2020	Turtle	P	*S.* Typhimurium	35	11	00	[22]
	2019	Turtle	P	*S.* Oranienburg	26	08	00	[22]
	2017	Turtles	P	*S.* Agbeni	76	30	00	[22]
	2015	Crested Geckos	P	*S.* Muenchen	22	03	00	[22]
	2015	Turtle	P	*S.* Sandiego *S.* Poona	133	38	00	[22]
	2014	Crested Dragon	P	*S.* Cotham	166	61	00	[22]
	2014	Snake	P	*S.* Enteritidis	1*	1	00	[23]
	2012–2013	Turtle	P	*S.* Sandiego *S.* Pomona *S.* Poona	473	78	00	[22]
	2009	Bearded dragon	P	*Salmonella enterica* subsp. houtenae 6,7:z4,z24:–	19	NI	00	[24]
	2009	Bearded dragon	P	*S.* Rubislaw	01	01	00	[25]
	2009	Bearded dragon	P	*S.* Apapa	01	01	00	[26]
	2009	Bearded dragon	P	*S.* Pomona	01	NI	NI	[27]
	2008	Turtle	P	*S.* Abony 4,5: b: enx	01	01	00	[28]
	2008	Snake	P	*S.* enterica subesp. Arizonae 41: z4, z23: -	03	01	00	[29]
	2007	Terrapin	P	*S.* Pomona	01	NI	NI	[30]
	2006	Bearded dragon	P	*S.* Apapa	03	NI	NI	[26]
	2005–2008	Snakes	P	*S.* Paratyphi B biovar Java 4,5,12: b: 1,2 *S.* Morehead 30: i: 1,5 *S. enterica* subesp. *Diarizonae* 47: -: -	03	NI	NI	[29]
	2005	Turtle	P	*S.* Braenderup 6,7: e, h: e, n, z15	06	00	00	[29]
	2005	Turtle	P	*S.* Paratyphi B	01	NI	00	[31]
	2003	Turtle	P	*S.* Enteritidis	01	01	00	[32]
	2003	Snake	P	*S. enterica* subsp. *arizonae*	01	01	01 **	[33]
	2000	Water dragon	P	*S.* Rubislaw	02	01	01 ***	[34]
	2000	Iguana	P	*Salmonella bongori* sorovar 44: Z23	01	01	00	[35]
	2000	Iguana	P	*S.* Poona	01	NI	00	[36]

Class	Year	Animal	Pet (P) Free (F)	*Salmonella* serovar	Cases: numbers of Illnesses (I); Hospitalizations (H); Deaths (D)			Ref.
					I	H	D	
Amphibians	2011	Frog	P	*S.* Typhimurium	241	72	00	[22]
	2009	African dwarf frog	P	*S.* Typhimurium	85	00	00	[37]
	2001	Frog and toad	NI	*S.* Javiana	55	09	00	[38]
Small Mammals	2020	Hedgehog	P	*S.* Typhimurium	49	11	00	[22]
	2019	Hedgehog	P	*S.* Typhimurium	54	08	00	[22]
	2018	Guinea Pig	P	*S.* Enteritidis	09	01	00	[22]
	2014	Frozen Feeder Rodents	****	*S.* Typhimurium	41	06	00	[22]
	2012	Hedgehog	P	*S.* Typhimurium	26	08	01	[22]
	2010	Frozen Feeder Rodents	****	*S. entérica* subsp. *enterica* 4,[5],12:i:-	34	01	00	[22]
	2008–2009	Feeder mice	****	*S.* Typhimurium DT191	12	NI	00	[39]
	2005–2006	Frozen Feeder Rodents	****	*S.* Typhimurium	04	00	00	[40]
	2003–2004	Rodent	P	*S.* Typhimurium	28	06	00	[41]
	2000	Hedgehog	F	*S.* Typhimurium	37	00	00	[42]
Wild Birds	2021	Wild Songbird	P / F	*Salmonella* spp.	29	14	00	[22]
	2001	Owl	NI	*S.* Typhimurium	40	04	00	[43]

NI: not informed.
**4-day-old neonate developed Salmonella meningitis.*
***3-month-old child with microcephaly.*
****3-week-old baby developed Salmonella meningitis and died.*
*****Used to feed pet reptiles.*

Table 2.
Salmonellosis outbreaks in humans associated with wild animals (2000–2021).

Cases of salmonellosis in humans caused by contact with wild animals kept away from their natural habitat have been reported (**Table 2**). These animals are often the carriers of not only *Salmonella* strains, but of other pathogens, for which there are not always effective control measures [44].

As shown in **Table 2**, among wild animals in captivity, reptiles cause most outbreaks of salmonellosis in humans [45, 46]. Salmonellosis in reptiles usually occurs asymptomatically [47]. The animals shed the bacterium intermittently and the elimination of the pathogen may increase due to stress factors [48]. Moreover, it is difficult to diagnose even in the presence of clinical signs [47]. However, human infections arising from human-reptile interaction can lead to clinical conditions ranging from mild to severe enteric infections, hospitalizations, and even deaths, especially in children, the elderly, and people with comorbidities [45].

Human contamination by *Salmonella* spp. from reptiles can be direct or indirect through secretions and excretions [49]. In a study conducted in southwest England between 2010 and 2013, 27.4% (48/175) of children under the age of five who had some contact with reptiles tested positive for *Salmonella* spp. and hospital admission rates totaled 50% for children under 1 year of age [50]. In another study conducted between 2008 and 2009 in New Zealand with 378 cloacal swabs of 24 different exotic reptile species kept as pets, 11.4% tested positive for *Salmonella enterica* subsb. Enterica, with emphasis on the serovars Onderstepoort (30.2%), Thompson (20.9%), Potsdam (14%), Wangata (14%), Infantis (11.6%), and Eastbourne (2.3%), which can also cause infectious conditions in humans [51].

The participation of free-living wild reptiles in the epidemiology of *Salmonella* should also be stressed. In a park in Poland, 16 free-living road-killed snakes were analyzed and 87.5% were positive for *Salmonella* spp. [52]. Briones et al. [53] analyzed free-living wild reptiles in preserved areas in Spain and found that 41.4% tested positive for *Salmonella enterica*, with 27 serotypes identified, 37.5% of which were associated with salmonellosis in humans. Regarding the group of affected animals, snakes and lizards are more prevalent than chelonians [51, 54].

A high prevalence of *Salmonella* spp. with serotype diversity is also found in amphibians. In a study conducted in Indiana County, Pennsylvania (USA), Chambers and Hulse [55] collected 92 free-living amphibians and found that 39.1% tested positive (23 salamanders and 13 frogs), with isolated serotypes Muenchen, Enteritidis, Typhimurium, Senftenberg, and Montevideo. The prevalence of *Salmonella* in amphibians was also examined in 58 Bufo marinus of the West Indies and 41% tested positive to five serotypes, especially *Salmonella enterica* subsp. Enterica serovar Javiana (33%) and *S.* Rubislaw (33%) [56]. In Thailand, eight serotypes of *Salmonella* spp. were identified (Hvittingfoss, Newport, Thompson, Stanley, Wandsworth, Panama, Muenchen, and subsp. diarizonae ser. 50:k:z) in 69.07% of the amphibians sampled in three different habitats - rural areas, protected areas, and urban areas. Of these serotypes, the first six have already been isolated in people in Thailand. Surprisingly, the animals coming from urban areas were negative [57]. The prevalence of *Salmonella* in amphibians regarding habitat remains unclear, although a possible cause is an environmental contamination by sewage [58]. This scenario is a public health concern because these amphibians can spread *Salmonella* spp. from the aquatic environment.

Outbreaks of salmonellosis in humans associated with contact with wild birds have been reported [59–61]. In 2000, an outbreak was reported in New Zealand caused by *S.* Typhimurium DT160, which led to the death of wild birds in rural areas, mainly sparrows, and enteric infections in humans [62]. In 2001, New Zealand reported an outbreak of human salmonellosis by *S.* Typhimurium DT160 related to contact with dead wild birds [63]. In 2001, two outbreaks were reported in the United States with at least 40 people contaminated with *S.* Typhimurium from the dissection of owls in two primary schools [43].

Between 1995 and 2003, Pennycott et al. [64] sampled 779 free-living wild birds in Great Britain and identified that the most prevalent serotype was *S.* Typhimurium. In Norway, *S.* Typhimurium variant O: 4,12 was identified in 96% of the isolates in a sample of 470 wild birds of 26 different species [44]. Despite the acute and chronic infection caused by *Salmonella*, in wild birds, it is asymptomatic [65]. During migration, the immune system can be affected by stress, as in the case of hunger, which may lead to a greater release of the pathogen by feces, contributing to even greater environmental contamination.

Regarding wild mammals, some species such as African pygmy, ferrets, hedgehogs, prairie dogs, primates, and sugar gliders are raised as pets [66], which can cause salmonellosis infections and outbreaks from direct human contact with carrier animals or indirectly due to access to or living in the same contaminated environments as these animals [49]. Two human outbreaks in Norway, caused by *S*. Typhimurium 4.5, 12:i: 1.2 associated with hedgehogs, were reported from August to October 1996 and from July to November 2000, with 28 confirmed cases and 37 confirmed cases, respectively. In both cases, hedgehogs were the only common source, with positivity rates of 39% and 41%, respectively for the outbreaks of 1996 and 2000 [42].

Free-living wild mammals can also be asymptomatic carriers of *Salmonella* spp.; however, the prevalence is usually lower than when these animals are bred in captivity. From 2002 to 2010, 2713 animals were sampled in Italy, a total of 1612 mammals (1222 canids, 221 mustelids, 100 rodents, 69 ungulates), resulting in 7.25% animals positive for *Salmonella* spp. (63 canids, 25 mustelids, 5 ungulates, 24 birds), with emphasis on the Typhimurium serotype [67].

Notably, urbanization causes the spread of zoonotic agents due to new ecological interactions [68], from changes in eating habits to changes in migration routes [69]. When wild animals have access to urban spaces or modified environmental areas, they also come into contact with waste produced by humans, such as garbage and sewage. Moreover, these spaces are a food source for these animals [70]. Due to ineffective waste management, contaminated environments can be the source of numerous pathogens and favor the spread of antimicrobial resistance genes [71].

4. Antimicrobial resistance in wild environments

Antibiotic resistance is a health threat for humans, animals, and the environment [72]. Regarding microorganisms, this resistance initially occurred in the absence of anthropogenic factors and without the clinical application of antibiotics [73]. Thus, this resistance can develop naturally from the ecological evolution of microorganisms, such as gene mutation, due to environmental pressure [74]. However, human factors have contributed to greater antimicrobial resistance with a direct impact on ecosystems [75].

The anthropization of forest areas favors the contact of wildlife with domestic animals and humans [76]. In this regard, resistance can be acquired through the consumption of water or food and can also occur through direct contact with human waste and sewage [77]. Another factor that favors the spread of resistant microorganisms is the displacement capacity of the carrier [78]. However, although wildlife has not had direct access to antibiotics, natural habitats altered by demographic expansion can enhance the sharing of resistance across different ecological niches [79]. According to Jechalke et al. [80], free-living wild animals that have not been exposed to antibiotics exhibit high drug resistance rates due to environmental contamination. Gilliver et al. [81] identified a marked prevalence of antibiotic-resistant wild rodents that were not exposed to antimicrobials.

Residues from antibiotics applied in human and veterinary medicine enable the spread of resistant agents to wild species through environmental contamination, especially among those that share the same habitat [82, 83]. Therefore, antimicrobial resistance can be greater in forest areas close to rural properties due to the inappropriate use of antibiotics to prevent and control diseases or due to their use as animal performance enhancers [84]. These conditions increase contamination of the environment, water resources, the food chain, and, finally, human and animal health. Sub-doses of antibiotics may select multiresistant plasmids [85]. It should be noted that resistance plasmids are highly associated with cases of resistance to beta-lactam antibiotics in gram-negative bacteria from extended-spectrum β-lactamases (ESBLs) [86].

Antimicrobial residues that accumulate in sediments can determine changes in the microbiome of soils in aquatic and terrestrial environments [87]. These effects are intensified by erosion, surface runoff, and displacement of soil minerals [88]. When these elements reach the springs or are used for irrigation, or when the sediment is used as decomposed organic matter for agriculture, cyclic, rotational maintenance of this contamination occurs in the environment [89].

5. Conclusion

The alteration of forest areas through anthropic actions favors increases the spread of infectious agents since it enables a pathogen to leave its ecosystem and natural hosts and adapt to other environments and reservoirs. These new interactions create different environmental, epidemiological, and sanitary patterns, especially in emerging and neglected zoonoses, and hinder control and eradication, as in the case of salmonellosis. Wild animals raised as pets or illegally kept in captivity also increase the prevalence of salmonellosis cases in humans mainly caused by exotic serotypes of *Salmonella*, due to direct contact with the bacterial strains in these animals.

Since *Salmonella* spp. can also be transmitted by wild animals, prevention and control measures should include sanitary-environmental factors and an international health inspection system for emerging and re-emerging diseases originating from wild fauna. These measures would enable a better understanding of the epidemiology and pathogenesis of infections and reduce economic and health costs with diagnosis and medications.

Acknowledgements

The authors thank Universidade Estadual de Santa Cruz for the financial support.

Conflict of interest

The authors declare no conflict of interest.

DOI: http://dx.doi.org/10.5772/intechopen.102618

Author details

Eliege Jullia Eudoxia dos Santos[1], Amanda Teixeira Sampaio Lopes[1] and Bianca Mendes Maciel[1,2]*

1 Graduation Program in Animal Science, Santa Cruz State University, Ilhéus, Bahia, Brazil

2 Department of Biological Sciences, Santa Cruz State University, Ilhéus, Bahia, Brazil

*Address all correspondence to: bmmaciel@uesc.br

References

[1] Kruse H, Kirkemo A-M, Handeland K. Wildlife as source of zoonotic infections. Emerging Infectious Diseases. 2004;**10**: 2067-2072. DOI: 10.3201/eid1012.040707

[2] Maciel BM, Rezende RP. Sriranganathan N. Salmonella enterica: Latency. In: Mares M, editor. Current Topics in *Salmonella* and Salmonellosis. Rijeka: InTech; 2017. pp. 41-58. DOI: 10.5772/67173

[3] Maciel BM, Argôlo Filho RC, Nogueira SSC, Dias JCT, Rezende RP. High prevalence of *Salmonella* in tegu lizards (*Tupinambis merianae*), and susceptibility of the serotypes to antibiotics. Zoonoses and Public Health. 2009;**57**:e26-e32. DOI: 10.1111/j.1863-2378.2009.01283.x

[4] Argolo LS, Maciel BM, Argolo Filho RC, Rezende RP. Detection of Salmonella in captive snakes by traditional isolation and specific PCR. Agrotrópica. 2013;**25**:215-222. DOI: 10.21757/0103-3816.2013v25 n3p215-222

[5] Santos EJE, Azevedo RP, Lopes ATS, Rocha JM, Albuquerque GR, Wenceslau AA, et al. *Salmonella* spp. in wild free-living birds from Atlantic Forest fragments in southern Bahia, Brazil. BioMed Research International. 2020;**2020**:1-7. DOI: 10.1155/2020/7594136

[6] Brasil. Ministério da Saúde. Secretaria de Vigilância em Saúde. Manual técnico de diagnóstico laboratorial de Salmonella spp.: diagnóstico laboratorial do gênero *Salmonella*/Ministério da Saúde. Série A. Normas e Manuais Técnicos. Secretaria de Vigilância em Saúde, Fundação Oswaldo Cruz. Laboratório de Referência Nacional de Enteroinfecções Bacterianas, Instituto Adolfo Lutz. – Brasília-DF: Ministério da Saúde (MS). 2011. 60. ISBN 978-85-334-1792-2

[7] Ricke SC, Ok KK, Foley S, Nayak R. *Salmonella*. In: Labbé RG, García S, editors. Guide to Foodborne Pathogens. 2nd ed. New Jersey (USA): Wiley-Blackwell; 2013. pp. 112-137. DOI: 10.1002/9781118684856.ch7

[8] Grimont PA, Weill FX. Antigenic formulae of the *Salmonella* serovars. WHO Collaborating Centre for Reference and Research on Salmonella. 2007;**9**:1-166

[9] Rodrigues DP. Perspectivas atuais e falhas no diagnóstico antigênico de Salmonella spp.: importância no reconhecimento dos sorovares circulantes, emergentes e exóticos. In: Memoria Del Seminário Internacional Sobre Salmonelose Aviar. Rio de Janeiro (Brazil), 1 CD-ROM; 2011. pp. 1-7

[10] Guibourdenche M, Roggentin P, Mikoleit M, Fields PI, Bockemühl J, Grimont PA, et al. Supplement 2003-2007 (No. 47) to the white-Kauffmann-Le minor scheme. Research in Microbiology. 2010;**161**:26-29. DOI: 10.1016/j.resmic.2009.10.002

[11] Lamas A, Miranda JM, Regal P, Vázquez B, Franco CM, Cepeda A. A comprehensive review of non-enterica subspecies of *Salmonella enterica*. Microbiological Research. 2018;**206**:60-73. DOI: 10.1016/j.micres.2017.09.010

[12] Gortazar C, Diez-Delgado I, Barasona JA, Vicente J, De La Fuente J, Boadella M. The wild side of disease control at the wildlife-livestock-human interface: A review. Frontiers in Veterinary Science. 2015;**1**:1-12. DOI: 10.3389/fvets.2014.00027

[13] Hirsh DC, Zee YC. Microbiologia Veterinária. 1st ed. Rio De Janeiro: Guanabara Koogan; 2003. p. 446

[14] Brenner FW, Villar RG, Angulo FJ, Tauxe R, Swaminathan B. *Salmonella* nomenclature. Journal of Clinical Microbiology. 2000;**38**:2465-2467. DOI: 10.1128/JCM.38.7.2465-2467.2000

[15] Fookes M, Schroeder GN, Langridge GC, Blondel CJ, Mammina C, Connor TR, et al. *Salmonella* bongori frovides insights into the evolution of the salmonellae. PLoS Pathogens. 2011;7:e1002191. DOI: 10.1371/journal.ppat.1002191

[16] House JK, Mainar-Jaime RC, Smith BP, House AM, Kamiya DY. Risk factors for nosocomial *Salmonella* infection among hospitalized horses. Journal of the American Veterinary Medical Association. 1999;**214**:1511-1516

[17] Strawn LK, Fortes ED, Bihn EA, Nightingale KK, Gröhn YT, Worobo RW, et al. Landscape and meteorological factors affecting prevalence of three food-borne pathogens in fruit and vegetable farms. Applied and Environmental Microbiology. 2013;**79**:588-600. DOI: 10.1128/aem.02491-12

[18] Gazal LE, De Brito KC, Cavalli LS, Kobayashi RK, Nakazato G, Otutumi LK, et al. *Salmonella* sp. em peixes – qual a importância para sanidade em pescado? Pesquisa Agropecuária Gaúcha. 2018;**24**:55-64. DOI: 10.36812/pag.2018241/255-64

[19] Fish JT, Pettibone GW. Influence of freshwater sediment on the survival of *Escherichia coli* and *Salmonella* sp. as measured by three methods of enumeration. Letters in Applied Microbiology. 1995;**20**:277-281. DOI: 10.1111/j.1472-765x.1995.tb00445.x

[20] Lim CH, Flint KP. The effects of nutrients on the survival of *Escherichia coli* in lake water. Journal of Applied Bacteriology. 1989;**66**:559-569. DOI: 10.1111/j.1365-2672.1989.tb04578.x

[21] Schloegel LM, Daszak P, Nava A. Medicina da conservação: buscando causas e soluções práticas para doenças infecciosas emergentes. Natureza & Conservação. 2005;**3**:29-41

[22] Center for Desease ontrol and Prevention (CDC). US Outbreaks of Zoonotic Diseases Spread between Animals & People. Healthy Pets, Healthy People. CDC. 2021. Available from: https://www.cdc.gov/healthypets/outbreaks.html [Accessed: 09-28-2021]

[23] Glick SR, Alter SJ. *Salmonella* enterica serotype Enteritidis meningitis in a 4-day-old neonate. Infectious Diseases in Clinical Practice. 2014;**22**:46-48. DOI: 10.1097/ipc.0b013e318280d927

[24] Lowther SA, Medus C, Scheftel J, Leano F, Jawahir S, Smith K. Foodborne outbreak of *Salmonella* subspecies IV infections associated with contamination from bearded dragons. Zoonoses and Public Health. 2011;**58**:560-566. DOI: 10.1111/j.1863-2378.2011.01403.x

[25] Moffatt CR, Lafferty AR, Khan S, Krsteski R, Valcanis M, Powling J, et al. *Salmonella* Rubislaw gastroenteritis linked to a pet lizard. Medical Journal of Australia. 2010;**193**:54-55. DOI: 10.5694/j.1326-5377.2010.tb03743.x

[26] Cooke FJ, De Pinna E, Maguire C, Guha S, Pickard DJ, Farrington M, et al. First report of human infection with *Salmonella* enterica serovar Apapa resulting from exposure to a pet lizard. Journal of Clinical Microbiology. 2009;**47**:2672-2674. DOI: 10.1128/jcm.02475-08

[27] Böhme H, Fruth A, Rebmann F, Sontheimer D, Rabsch W.

Reptile-associated salmonellosis in a breastfed infant. Klinische Padiatrie. 2009;**221**:74-75. DOI: 10.1055/s-0028-1112157

[28] Van Meervenne E, Botteldoorn N, Lokietek S, Vatlet M, Cupa A, Naranjo M, et al. Turtle-associated *Salmonella* septicaemia and meningitis in a 2-month-old baby. Journal of Medical Microbiology. 2009;**58**:1379-1381. DOI: 10.1099/jmm.0.012146-0

[29] Bertrand S, Rimhanen-Finne R, Weill FX, Rabsch W, Thornton L, Perevoščikovs J, et al. *Salmonella* infections associated with reptiles: The current situation in Europe. Eurosurveillance. 2008;**13**:18902. DOI: 10.2807/ese.13.24.18902-en

[30] O'Byrne AM, Mahon M. Reptile-associated salmonellosis in residents in the south east of Ireland 2005-2007. Eurosurveillance. 2008;**13**:18830. DOI: 10.2807/ese.13.15.18830-en

[31] Nagano N, Oana S, Nagano Y, Arakawa Y. A severe *Salmonella* enterica serotype Paratyphi B infection in a child related to a pet turtle, Trachemys scripta elegans. Japanese Journal of Infectious Diseases. 2006;**59**:132-134

[32] Stam F, Romkens TEH, Hekker TAM, Smulders YM. Turtle-associated human salmonellosis. Clinical Infectious Diseases. 2003;**37**:e167-e169. DOI: 10.1086/379612

[33] Mahajan RK, Khan SA, Chandel DS, Kumar N, Hans C, Chaudhry R. Fatal case of Salmonella enterica subsp. arizonae gastroenteritis in an infant with microcephaly. Journal of Clinical Microbiology. 2003;**41**:5830-5832. DOI: 10.1128/jcm.41.12.5830-5832.2003

[34] Ward L. Fatal neonatal *Salmonella* Rubislaw infection in household with pet reptile in England. Weekly releases (1997-2007). 2000;**4**:1663. DOI: 10.2807/esw.04.06.01663-en

[35] Nowinski RJ, Albert MC. *Salmonella* osteomyelitis secondary to iguana exposure. Clinical Orthopaedics and Related Research. 2000;**372**:250-253. DOI: 10.1097/00003086-200003000-00027

[36] Sam WIC, Mackay AD. *Salmonella* meningitis and a green iguana. Journal of the Royal Society of Medicine. 2000;**93**:318-319. DOI: 10.1177/014107680009300613

[37] Centers for Disease Control and Prevention (CDC). Multistate outbreak of human Salmonella Typhimurium infections associated with aquatic frogs-United States, 2009. MMWR. Morbidity and Mortality Weekly Report. 2010;**58**:1433-1436

[38] Srikantiah P, Lay JC, Hand S, Crump JA, Campbell J, Van Duyne MS, et al. *Salmonella* enterica serotype Javiana infections associated with amphibian contact, Mississippi, 2001. Epidemiology and Infection. 2004;**132**:273-281. DOI: 10.1017/s0950268803001638

[39] Harker KS, Lane C, De Pinna E, Adak GK. An outbreak of *Salmonella* Typhimurium DT191a associated with reptile feeder mice. Epidemiology and Infection. 2011;**139**:1254-1261. DOI: 10.1017/S0950268810002281

[40] Fuller CC, Jawahir SL, Leano FT, Bidol SA, Signs K, Davis C, et al. A multistate *Salmonella* Typhimurium outbreak associated with frozen vacuum-packed rodents used to feed snakes. Zoonoses and Public Health. 2008;**55**:481-487. DOI: 10.1111/j.1863-2378.2008.01118.x

[41] Swanson SJ, Snider C, Braden CR, Boxrud D, Wünschmann A, Rudroff JA,

et al. Multidrug-resistant *Salmonella* enterica serotype Typhimurium associated with pet rodents. New England Journal of Medicine. 2007;**356**:21-28. DOI: 10.1056/nejmoa060465

[42] Handeland K, Refsum T, Johansen BS, Holstad G, Knutsen G, Solberg I, et al. Prevalence of *Salmonella* Typhimurium infection in Norwegian hedgehog populations associated with two human disease outbreaks. Epidemiology and Infection. 2002;**128**:523-527. DOI: 10.1017/s0950268802007021

[43] Smith KE, Anderson F, Medus C, Leano F, Adams J. Outbreaks of salmonellosis at elementary schools associated with dissection of owl pellets. Vector Borne and Zoonotic Diseases. 2005;**5**:133-136. DOI: 10.1089/vbz.2005.5.133

[44] Refsum T, Handeland K, Baggesen DL, Holstad G, Kapperud G. Salmonellae in avian wildlife in Norway from 1969 to 2000. Applied and Environmental Microbiology. 2002;**68**:5595-5599. DOI: 10.1128/aem.68.11.5595-5599.2002

[45] Mermin J, Hutwagner L, Vugia D, Shallow S, Daily P, Bender J, et al. Reptiles, amphibians, and human *Salmonella* infection: A population-based, case–control study. Clinical Infectious Diseases. 2004;**38**:S253-S261. DOI: 10.1086/381594

[46] Wells EV, Boulton M, Hall W, Bidol SA. Reptile-associated salmonellosis in preschool-aged children in Michigan, January 2001-June 2003. Clinical Infectious Diseases. 2004;**39**:687-691. DOI: 10.1086/423002

[47] Bemis DA, Grupka LM, Liamthong S, Folland DW, Sykes JM, Ramsay EC. Clonal relatedness of *Salmonella* isolates associated with invasive infections in captive and wild-caught rattlesnakes. Veterinary Microbiology. 2007;**120**:300-307. DOI: 10.1016/j.vetmic.2006.10.028

[48] Scheelings TF, Lightfoot D, Holz P. Prevalence of *Salmonella* in Australian reptiles. Journal of Wildlife Diseases. 2011;**47**:1-11. DOI: 10.7589/0090-3558-47.1.1

[49] Hale CR, Scallan E, Cronquist AB, Dunn J, Smith K, Robinson T, et al. Estimates of enteric illness attributable to contact with animals and their environments in the United States. Clinical Infectious Diseases. 2012;**54**: S472-S479. DOI: 10.1093/cid/cis051

[50] Murphy D, Oshin F. Reptile-associated salmonellosis in children aged under 5 years in south West England. Archives of Disease in Childhood. 2015;**100**:364-365. DOI: 10.1136/archdischild-2014-306134

[51] Kikillus K, Gartrell B, Motion E. Prevalence of *Salmonella* spp., and serovars isolated from captive exotic reptiles in New Zealand. New Zealand Veterinary Journal. 2011;**59**:174-178. DOI: 10.1080/00480169.2011.579246

[52] Zając M, Wasyl D, Różycki M, Bilska-Zając E, Fafiński Z, Iwaniak W, et al. Free-living snakes as a source and possible vector of *Salmonella* spp. and parasites. European Journal of Wildlife Research. 2016;**62**:161-166. DOI: 10.1007/s10344-016-0988-y

[53] Briones V, Tellez S, Goyache J, Ballesteros C, del Pilar LM, Dominguez L, et al. *Salmonella* diversity associated with wild reptiles and amphibians in Spain. Environmental Microbiology. 2004;**6**:868-871. DOI: 10.1111/j.1462-2920.2004.00631.x

[54] Marin C, Lorenzo-Rebenaque L, Laso O, Villora-Gonzalez J,

Vega S. Pet reptiles: A potential source of transmission of multidrug-resistant *Salmonella*. Frontiers in Veterinary Science. 2021;7:1157. DOI: 10.3389/fvets.2020.613718

[55] Chambers DL, Hulse AC. *Salmonella* serovars in the herpetofauna of Indiana County, Pennsylvania. Applied and Environmental Microbiology. 2006;**72**:3771-3773. DOI: 10.1128/aem.72.5.3771-3773.2006

[56] Drake M, Amadi V, Zieger U, Johnson R, Hariharan H. Prevalence of *Salmonella* spp. in cane toads (*Bufo marinus*) from Grenada, West Indies, and their antimicrobial susceptibility. Zoonoses and Public Health. 2013;**60**:437-441. DOI: 10.1111/zph.12018

[57] Ribas A, Poonlaphdecha S. Wild-caught and farm-reared amphibians are important reservoirs of *Salmonella*, a study in north-East Thailand. Zoonoses and Public Health. 2017;**64**:106-110. DOI: 10.1111/zph.12286

[58] Taylor SK, Green DE. Bacterial diseases. In: Wright KM, Whitaker BR, editors. Amphibian Medicine and Captive Husbandry. 1st ed. Melbourne: Krieger Publishing Company; 2001. pp. 159-179

[59] Bloomfield SJ, Benschop J, Biggs PJ, Marshall JC, Hayman DT, Carter PE, et al. Genomic analysis of *Salmonella* enterica serovar Typhimurium DT160 associated with a 14-year outbreak, New Zealand, 1998-2012. Emerging Infectious Diseases. 2017;**23**:906-913. DOI: 10.3201/eid2306.161934

[60] Lawson B, de Pinna E, Horton RA, Macgregor SK, John SK, Chantrey J, et al. Epidemiological evidence that garden birds are a source of human salmonellosis in England and Wales. PLoS One. 2014;**9**:e88968. DOI: 10.1371/journal.pone.0088968

[61] Mather AE, Lawson B, de Pinna E, Wigley P, Parkhill J, Thomson NR, et al. Genomic analysis of *Salmonella* enterica serovar Typhimurium from wild passerines in England and Wales. Applied and Environmental Microbiology. 2016;**82**:6728-6735. DOI: 10.1128/aem.01660-16

[62] Alley M, Connolly J, Fenwick S, Mackereth G, Leyland M, Rogers L, et al. An epidemic of salmonellosis caused by *Salmonella* Typhimurium DT160 in wild birds and humans in New Zealand. New Zealand Veterinary Journal. 2002;**50**:170-176. DOI: 10.1080/00480169.2002.36306

[63] Thornley CN, Simmons GC, Callaghan ML, Nicol CM, Baker MG, Gilmore KS, et al. First incursion of *Salmonella* enterica serotype Typhimurium DT160 into New Zealand. Emerging Infectious Diseases. 2003;**9**:493-495. DOI: 10.3201/eid0904.020439

[64] Pennycott TW, Park A, Mather HA. Isolation of different serovars of *Salmonella* enterica from wild birds in Great Britain between 1995 and 2003. Veterinary Record. 2006;**158**:817-820. DOI: 10.1136/vr.158.24.817

[65] Troxler S, Hess C, Konicek C, Knotek Z, Barták P, Hess M. Microdilution testing reveals considerable and diverse antimicrobial resistance of *Escherichia coli*, thermophilic *Campylobacter* spp. and *Salmonella* spp. isolated from wild birds present in urban areas. European Journal of Wildlife Research. 2017;**63**:1-11. DOI: 10.1007/s10344-017-1125-2

[66] Reaser JK, Clark EE, Meyers NM. All creatures great and minute: A public policy primer for companion animal zoonoses. Zoonoses and Public Health. 2008;**55**:385-401. DOI: 10.1111/j.1863-2378.2008.01123.x

[67] Botti V, Navillod FV, Domenis L, Orusa R, Pepe E, Robetto S, et al. *Salmonella* spp. and antibiotic-resistant strains in wild mammals and birds in North-Western Italy from 2002 to 2010. Veterinaria Italiana. 2013;**49**:195-202. DOI: 10.12834/VetIt.2013.492.201.208

[68] Hoornweg D, Bhada-Tata P. What a Waste: A Global Review of Solid Waste Management. Urban Development Series Knowledge Papers no. 15. Washington, DC: World Bank; 2012. p. 98. Available from: https://openknowledge.worldbank.org/handle/10986/17388 License: CC BY 3.0 IGO [Accessed: 09-28-2021]

[69] Miranda AC. Mechanisms of behavioral change in urban animals: The role of microevolution and phenotypic plasticity. In: Murgui E, Hedblom M, editors. Ecology and Conservation of Birds in Urban Environments. Cham: Springer International Publishing; 2017. pp. 113-132. DOI: 10.1007/978-3-319-43314-1_7

[70] Plaza PI, Lambertucci SA. How are garbage dumps impacting vertebrate demography, health, and conservation? Global Ecology and Conservation. 2017;**12**:9-20. DOI: 10.1016/j.gecco.2017.08.002

[71] Camacho M, Hernández JM, Lima-Barbero JF, Höfle U. Use of wildlife rehabilitation centres in pathogen surveillance: A case study in white storks (*Ciconia ciconia*). Preventive Veterinary Medicine. 2016;**130**:106-111. DOI: 10.1016/j.prevetmed.2016.06.012

[72] Larsson DG, Andremont A, Bengtsson-Palme J, Brandt KK, de Roda Husman AM, Fagerstedt P, et al. Critical knowledge gaps and research needs related to the environmental dimensions of antibiotic resistance. Environment International. 2018;**117**:132-138. DOI: 10.1016/j.envint.2018.04.041

[73] D'Costa VM, King CE, Kalan L, Morar M, Sung WW, Schwarz C, et al. Antibiotic resistance is ancient. Nature. 2011;**477**:457-461. DOI: 10.1038/nature10388

[74] Wellington EM, Boxall AB, Cross P, Feil EJ, Gaze WH, Hawkey PM, et al. The role of the natural environment in the emergence of antibiotic resistance in gram-negative bacteria. The Lancet Infectious Diseases. 2013;**13**:155-165. DOI: 10.1016/s1473-3099(12)70317-1

[75] Karkman A, Pärnänen K, Larsson DG. Fecal pollution can explain antibiotic resistance gene abundances in anthropogenically impacted environments. Nature Communications. 2019;**10**:1-8. DOI: 10.1038/s41467-018-07992-3

[76] Kukielka E, Barasona JA, Cowie CE, Drewe JA, Gortazar C, Cotarelo I, et al. Spatial and temporal interactions between livestock and wildlife in south Central Spain assessed by camera traps. Preventive Veterinary Medicine. 2013;**112**:213-221. DOI: 10.1016/j.prevetmed.2013.08.008

[77] Radhouani H, Silva N, Poeta P, Torres C, Correia S, Igrejas G. Potential impact of antimicrobial resistance in wildlife, environment and human health. Frontiers in Microbiology. 2014;**5**:23. DOI: 10.3389/fmicb.2014.00023

[78] Plaza-Rodríguez C, Alt K, Grobbel M, Hammerl JA, Irrgang A, Szabo I, et al. Wildlife as sentinels of antimicrobial resistance in Germany? Frontiers in Veterinary Science. 2021;**7**:1251. DOI: 10.3389/fvets.2020.627821

[79] Jones KE, Patel NG, Levy MA, Storeygard A, Balk D, Gittleman JL, et al. Global trends in emerging infectious diseases. Nature. 2008;**451**:990-993. DOI: 10.1038/nature06536

[80] Jechalke S, Heuer H, Siemens J, Amelung W, Smalla K. Fate and effects of veterinary antibiotics in soil. Trends in Microbiology. 2014;**22**:536-545. DOI: 10.1016/j.tim.2014.05.005

[81] Gilliver MA, Bennett M, Begon M, Hazel SM, Hart CA. Antibiotic resistance found in wild rodents. Nature. 1999;**401**:233-234. DOI: 10.1038/45724

[82] Friese A, Schulz J, Hoehle L, Fetsch A, Tenhagen BA, Hartung J, et al. Occurrence of MRSA in air and housing environment of pig barns. Veterinary Microbiology. 2012;**158**:129-135. DOI: 10.1016/j.vetmic.2012.01.019

[83] Huijbers PM, Flach CF, Larsson DG. A conceptual framework for the environmental surveillance of antibiotics and antibiotic resistance. Environment International. 2019;**130**:104880. DOI: 10.1016/j.envint.2019.05.074

[84] Thomas M, Fenske GJ, Antony L, Ghimire S, Welsh R, Ramachandran A, et al. Whole genome sequencing-based detection of antimicrobial resistance and virulence in non-typhoidal *Salmonella enterica* isolated from wildlife. Gut Pathogens. 2017;**9**:1-9. DOI: 10.1186/s13099-017-0213-x

[85] Gullberg E, Albrecht LM, Karlsson C, Sandegren L, Andersson DI. Selection of a multidrug resistance plasmid by sublethal levels of antibiotics and heavy metals. MBio. 2014;**5**:e01918-e01914. DOI: 10.1128/mbio.01918-14

[86] D'Andrea MM, Arena F, Pallecchi L, Rossolini GM. CTX-M-type β-lactamases: A successful story of antibiotic resistance. International Journal of Medical Microbiology. 2013;**303**:305-317. DOI: 10.1016/j.ijmm.2013.02.008

[87] Boxall AB, Kolpin DW, Halling-Sørensen B, Tolls J. Peer reviewed: Are veterinary medicines causing environmental risks? Environmental Science & Technology. 2003;**37**:286A-294A. DOI: 10.1021/es032519b

[88] Hirsch R, Ternes T, Haberer K, Kratz KL. Occurrence of antibiotics in the aquatic environment. Science of the Total Environment. 1999;**225**:109-118. DOI: 10.1016/s0048-9697(98)00337-4

[89] Kim S, Aga DS. Potential ecological and human health impacts of antibiotics and antibiotic-resistant Bacteria from wastewater treatment plants. Journal of Toxicology and Environmental Health, Part B. 2007;**10**:559-573. DOI: 10.1080/15287390600975137

Chapter 5

Salmonella: The Critical Enteric Foodborne Pathogen

Mohd Afendy Abdul Talib, Son Radu, Cheah Yoke Kqueen and Farinazleen Mohamad Ghazali

Abstract

Persistent cases of *Salmonella* infection have urged great attention and surveillance on this foodborne pathogen. *Salmonella* continues to be a significant foodborne disease worldwide for both animals and people in the twenty-first century. It is one of the leading causes of foodborne pathogens infecting animals and humans. Salmonellosis is a principal cause of food poisoning and is, hence, a severe public health problem. The history, classification and nomenclature of *Salmonella*, as well as its characteristics, clinical manifestations, epidemiology and route of contamination, will be covered in this chapter to help readers gain a better understanding and overview of this microbe.

Keywords: *Salmonella*, foodborne, pathogens, food poisoning

1. Introduction

Foodborne illnesses are defined by World Health Organization (WHO) as diseases, usually either infectious or toxic in nature, caused by agents that enter the body through the ingestion of food. Foodborne diseases could be caused by a wide range of biological and chemical agents or hazards resulting in varying degrees of severity, ranging from mild indisposition to chronic or life-threatening illness, or both. These agents include bacteria, viruses, protozoa, helminthes, and natural toxins, as well as chemical and environmental contaminants.

Foodborne illness or disease caused by foodborne pathogens occurred every year in both developed and developing countries throughout the world. The incidence of foodborne disease is difficult to be estimated globally but it was reported that an estimate of 600 million or almost 1 in 10 people in the world fall ill after eating contaminated food and 420,000 die every year. Centers for Disease Control & Prevention, US in 2011 [1] estimated that roughly one of six Americans or 48 million people get sick, 128,000 are hospitalized, and 3,000 die of foodborne diseases. Among these, children under 5 years of age carry 40% of the foodborne disease burden, with 125,000 deaths occuring every year [2].

2. History background

Salmonella was a prevalent pathogen that infected the digestive tracts of both human and animals. *Salmonella* contamination in food, water and the natural environment is mainly caused by faecal contamination in the environment (**Figure 1**). Some *Salmonella* serovars are host-specific; for instance, serovars Typhi and Paratyphi A can only be colonised in humans, serovar Abortusovis in sheep and serovar Gallinarum in fowl. *Salmonella* serovars also induce infectious syndromes distinct to their type; for example, the highly adapted serovar Typhi causes systemic infection called typhoid exclusively in humans. Serovar Typhimurium causes non-typhoidal salmonellosis (NTS) in human, which is one of the most prevalent serotypes responsible for infections, including acute gastroenteritis in humans [1] and animal species like hens [4], pig [5] and mice [6]. Serovar Abortusovis causes high rates of abortion in flocks, ewes, sheep and goat [7, 8], and serovar Dublin originally discovered in cattle, which adapted to infect other animals such as bovines and fox [9–11]. Infection by serovar Dublin in human is rare but causes rather severe invasive bloodstream infection [10].

In 1880, Karl Eberth discovered a bacillus-like pathogen in the spleen and Peyer's patches of typhoid patients. He was a student of the famous Rudolf Virchow [12]. Four years later (in 1884), Georg Gaffky, a German microbiologist, successfully grew the pure culture of the bacterium [13]. Theobald Smith was the first to discover what would be later known as *Salmonella enterica* (var. Choleraesuis) while working in the Veterinary Division of the United States Department of Agriculture (USDA) as a research laboratory assistant in a department headed by Daniel Elmer Salmon, the veterinary pathologist. At first, the agent responsible for swine fever or hog cholera was thought to be caused by *Salmonella* Choleraesuis, prompting both Salmon and Smith to name the bacterium "Hog-cholerabacillus" [14]. In fact, Salmon and Smith were first to discover and isolate *S.* Choleraesuis from pigs in 1886. Incidentally, it was not until 1900 that the name and genus *Salmonella* was used when it was proposed by Joseph Leon Lignières and named after Daniel Elmer Salmon as an honorific attribute to the discovery made by his group [15].

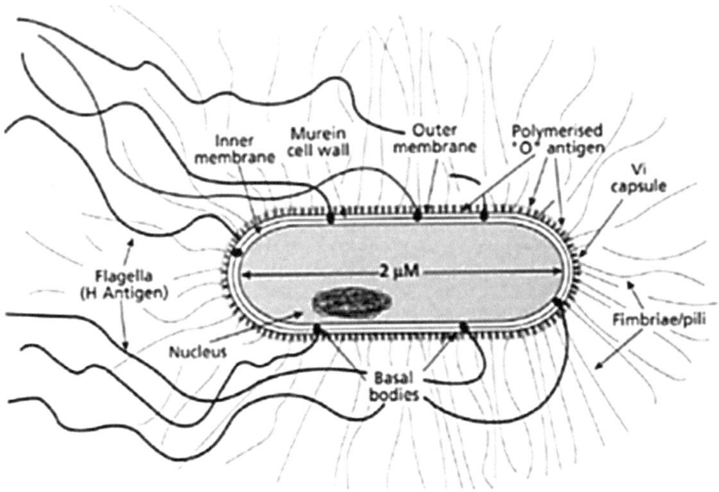

Figure 1.
Schematic diagram showing important structural components of Salmonella *Typhi (Source: Hu & Kopecko, 2003 [3]).*

3. Classification and nomenclature

To characterise and communicate about this bacterial genus, scientists have used comprehensive *Salmonella* nomenclature. Historically, *Salmonella* strains were classified based on their epidemiology, host range, clinical symptoms, biochemical reactions and surface antigenic patterns. Previously, the name *Salmonella* was derived from the geographical location where the first strain is isolated; for example, *S.* Heidelberg, *S.* Derby, *S.* London. Other than that, it was also named according to its clinical conditions or host specificity, such as *S. typhimurium, S. enteritidis, S. typhi,* 'S. gallinarum', 'S. abortusovis' or *S. choleraesuis*. However, it was soon realised that these so-called species were ubiquitous [16].

A great number of serovars have been described as a result of O and H antigens analysis initiated by White in 1926 and continued by Kauffmann later in 1941. The species that were defined by Kauffman as 'a group of related sero-fermentative phage types' created more than 2000 serovar names according to the species. However, the concept of one serovar one species was discovered to be unsuitable since biochemical test still could not separate most of the serovars. Although the terms serotype and serovar are interchangeable, according to the Bacteriological Code, serovar is now recommended for scientific communication (1990 Revision).

In the early years, Kauffman identified *Salmonella* serovars in 1966 based on its antigenic composition, and there were multiple species within its genus. He found *Salmonella* serovars and a variety of species within the genus using the antigenic formula. Some clinically important *Salmonella* strains were found before 1966, and the majority of serovars were named after the illness and/or the host, such as *S. typhi* and *S. typhimurium,* or by the geographical area or origin of the species that were first isolated.

The epidemiologic classification of *Salmonella* is based on the preferences of the hosts. *S.* Typhi was the first of the host-restricted serotypes that only infect humans. The second group includes host-adapted serotypes associated with one host species but can cause disease in other hosts. An example of a host-adapted serotype was *S.* Pullorum, which was discovered in an avian. The remaining serotypes are in the third group. Each year, the three most common serotypes recovered from humans are *Salmonella* Enteritidis, *Salmonella* Typhimurium, and *Salmonella* Heidelberg.

Given the complexities of the various *Salmonella* species, it was proposed that the genus *Salmonella* was separated into three species: *S. choleraesuis,* 'S. typhosa' (*S. typhi*) and 'S. kauffmannii'. 'S. kauffmannii' contains entirely additional serovars. *S. enterica*, according to Kauffmann and Edwards (1952), should include all *salmonellae. S. enterica* subsp. *enterica (S. enterica* subsp. I*)* are the most frequent *Salmonella* serovar among the approximately >2600 *Salmonella* serovars that have been found to date [17]. Human and warm-blooded animal infections with *Salmonella* infections are virtually always caused by strains belonging to the O-antigen serogroups (bacteria's surface of their outermost layer), A, B, C1, C2, D, and E [18]. The oligosaccharides associated with lipopolysaccharide determine the O antigen.

Salmonella infections that are typically isolated from cold-blooded animals and the natural environment but uncommonly isolated from human are caused by serovars in the *S. enterica* subspecies *salamae (S. enterica* subsp. II*)*, *arizonae (S. enterica* subsp. IIIa*)*, *diarizonae (S. enterica*subsp. IIIb), *houtenae (S. enterica* subsp. IV), *indica (S. enterica* subsp. VI), and *S. bongori*.

The White-Kauffmann-Le Minor scheme, previously designated as Kauffman-White scheme, described *Salmonella's* characterisation based on antibody recognition with antigens on *Salmonella's* surface. Three major antigenic determinants are used to classify *Salmonella* into three groups in the scheme: flagellar H antigens, somatic O antigens and virulence (Vi) capsular K antigens. This scheme is an established document that lists all identified serovars [19, 20]. The document has been updated by the World Health Organisation (WHO) Collaborating Centre for Reference and Research on *Salmonella* at the Pasteur Institute, Paris, France, and every newly identified serovar is reported in the journal *Research in Microbiology* yearly. The WHO Collaborating Centre for Reference and Research on *Salmonella*, in the Pasteur Institute in Paris, has updated the document. As a result, every newly discovered *Salmonella* serovar as well as other relevant microorganism is reported in the journal *Research in Microbiology* every year [21].

On the recommendation of the WHO collaborating centre, the current nomenclature used by the Centers for Disease Control and Prevention (CDC), Department of Health and Human Services, USA, is widely recognised. It is based on a two-species system (*S. enterica* and *S. bongori*), with multiple serovars in each species [22]. Microbiologists in clinical and public health have praised the system for meeting their needs [23]. *Salmonella* nomenclature is currently divided into two species, *S. enterica* and *S. bongori*, which has six subspecies and one subspecies, respectively. The nomenclature is summarised in **Table 1**. In addition, the relationship of phylogenetic tree among *Salmonella* subspecies is demonstrated in **Figure 2**.

Serovar names should not be printed in italics because they are no longer considered species names. For example, *S. enteritidis* becomes *S. enterica* subsp. *enterica* serovar Enteritidis, or simply written as *Salmonella* serovar Enteritidis and can be shortened to *S.* Enteritidis. Only serovars of *S. enterica* subsp. *enterica* are given names associated with disease syndrome or host habitat, while others represent the geographical origin of the first isolate found. On the other hand, other subspecies' serovars are identified by their antigenic formula O:H.

Serovars designated by antigenic formulae include the following: (i) subspecies designation (subspecies I through VI); (ii) O (somatic) antigens separated by a comma if needed, followed by colon: (iii) H (flagellar) antigens (phase 1) separated by a colon and (iv) H antigens (phase 2, if present) (for example, *Salmonella* serotype II 39:z10:-) [25].

Genus (capitalised, italic)	**Species (not capitalised, italic)**	**Subspecies (symbol) (not capitalised, italic)**	**Serovar name (with examples) (capitalised, Roman)**
Salmonella	*enterica*	*enterica* (subspecies I)	Choleraesuis, Enteritidis, Paratyphi, Typhi, Typhimurium
		salamae (subspecies II)	9,46:z:z39
		arizonae (subspecies IIIa)	43:z29:-
		diarizonae (subspecies IIIb)	6,7:l,v:1,5,7
		houtenae (subspecies IV)	21:m,t:-
		indica (subspecies VI)	59:z36:-
Salmonella	*bongori*	(subspecies V)	13,22:z39:-

Adapted from Su and Chiu [24].

Table 1.
Salmonella *nomenclature.*

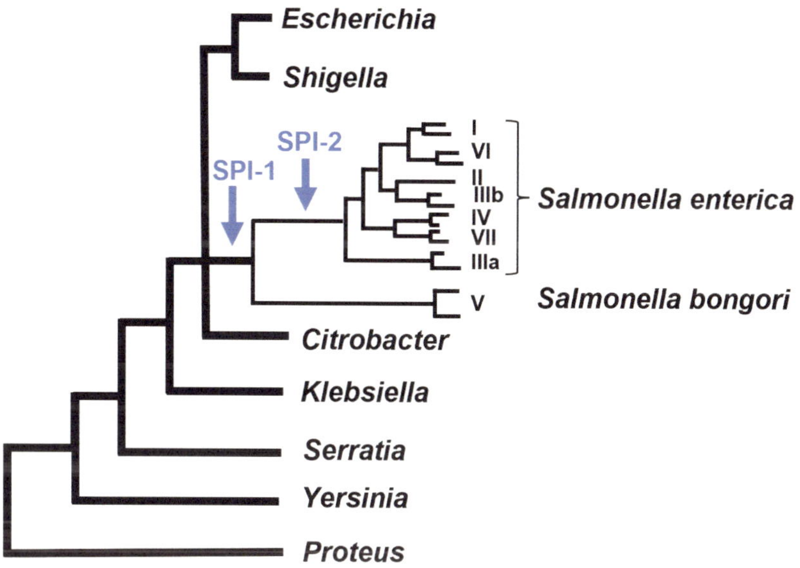

Figure 2.
Summary of relationship of phylogenetic tree among Salmonella *subspecies and other bacterial species (adapted from reference [1]).*

4. Characteristic

Salmonella belongs to the family *Enterobacteriaceae,* a gram-negative, facultative anaerobic and rod-shaped bacterium. The bacteria are 3–5 μm long and 0.7–1.5 μm wide. They are commonly motile with peritrichous flagella that help the bacteria to move, aerogenic, grow on nutrient agar, glucose-fermenting, non-lactose fermenting, often gas producer, urease-negative, citrate-utilising, oxidase-negative, potassium cyanide-negative and acetylmethyl carbinol-negative [26, 27].

Some serovars have peculiarities that are a mutant of normal motile serovars and can change to non-motile. The majority of isolates expressing H antigen exist in two phases: a motile phase I and a non-motile phase II. A Cragie tube can be used to switch non-motile cultures into the motile phase after they have been established in the primary culture [28]. Most *Salmonella* strains are prototrophic and can grow in a minimal medium utilising glucose as the sole carbon energy source and ammonium ion as a nitrogen source. Some host-adapted serovars (e.g., Typhi, Paratyphi A, Gallinarum, Sendai and Abortusovis) are auxotrophic and require one or more growth factors. The biochemical characteristic of *Salmonella* is shown in **Table 2**. Most species produce hydrogen sulphide, which can be detected by growing them on media containing ferrous sulphate, such as triple sugar iron (TSI). However, certain serovars, such as *S.* Typhi, never produce gas from glucose.

Salmonella lives predominantly in the intestines of animals and have adapted to live with their hosts [29]. *Salmonella enterica* subsp. *enterica* inhabits warm-blooded animals, whereas all other *S. enterica* subspecies and *S. bongori* live in cold-blooded animals and rarely infect humans. In terms of the types of hosts infected, *S. enterica* subsp. *enterica* serovars can be clustered to host-adapted, host-restricted and general-host [30]. **Table 3** lists out the host range of *Salmonella*. The host-adapted *Salmonella* infects habitually a single host but is capable of causing disease in another animal. Host-restricted *Salmonella* infects only a single host, while general-host *Salmonella*

Characteristics	***Salmonella enterica* subsp.**						***Salmonella bongori***
	enterica	***salamae***	***arizonae***	***diarizonae***	***houtenae***	***indica***	
Subspecies group	I	II	IIIa	IIIb	IV	VI	V
α-glutamyltransferase	d	+	–	+	+	+	+
β-Gluocuronidase	d	d	–	+	–	d	–
Dulcitol	+	+	–	–	–	d	+
Galacturonate	–	+	–	+	+	+	+
Gelatinase	–	+	+	+	+	+	–
Glucose	+	+	+	+	+	+	+
Hydrogen sulfide	+	+	+	+	+	+	+
Indole test	–	–	–	–	–	–	–
Lactose	–	–	–	+	–	+	d
Lysine decarboxylase	+	+	+	+	+	+	+
L(+)-tartrate	+	–	–	–	–	–	–
Malonate	–	+	+	+	–	–	–
Methyl red test	+	+	+	+	+	+	+
Murate	+	+	+	–	–	+	+
Ortho-nitrophenyl-β-*D*-Galactopyranoside test	–	–	+	+	–	d	+
Phage O1 susceptible	+	+	–	+	–	+	D
Potassium cyanide broth	–	–	–	–	–	–	–
Salicine	–	–	–	+	–	+	D
Sorbitol	+	+	+	+	+	–	+
Urease	–	–	–	–	–	–	–
Voger-Proskauer test	–	–	–	–	–	–	–

Note: +: more than 90% positive reactions; –: less than 10% positive reactions; d: 10–90% strains positive; ONPG: ortho-nitrophenyl-β-D-galactopyranoside.

Table 2.
Biochemical characteristics of Salmonella *species and subspecies.*

has the capability of infecting a variety of animals. However, the disease's progression may vary depending on the host [31].

Most *Salmonella* grows at a temperature of 7–48°C with the optimal growth at 37°C. However, some strains are capable of withstanding extremely low temperature, 2°C, or high temperature, 54°C [32]. Unfortunately, they are not commonly heat resistant and usually die within 1–10 min at 60°C and less than 1 min at 70°C.

The water activity (a_w) of foods influences the time and temperature needed to kill *Salmonella* and reduces the effectiveness of the heat treatment. *Salmonella* needs high water activity (a_w) between 0.94 and 0.99, optimally at 0.995 but can survive in foods with low a_w [33]. Low-a_w foods, such as nuts, flour, butter and chocolate, can extend the time and temperature required to kill the bacteria [34]. Some rare serotypes, such as *S.* Senftenberg strain 775 W, has 10–30 times more heat resistant than *S.* Typhimurium in low-a_w food products with high carbohydrate or high fat [35]. *Salmonella* grows at a pH value of 4–9 with the optimum growth at a pH value of 6.5–7.5 [36].

Classification	Serovar	Natural host	Rare hosts
Host restricted	Typhi	Humans	None
	Paratyphi A and C	Humans	None
	Sendai	Humans	None
	Abortusovis	Ovines	None
	Gallinarum	Poultry, birds	None
	Pullorum	Poultry, birds	None
	Typhisuis	Swine	None
	Abortusequi	Equines	None
Host adapted	Choleraesuis	Swine	Humans
	Dublin	Bovines	Human and bovines
General-host	Typhimurium	Humans, poultry, swine, bovines, and rodents	None
	Enteritidis	Humans, poultry, and rodents	Swine and bovines

Adapted from references [30, 31].

Table 3.
Host range of Salmonella *enterica subsp. enterica serovars.*

5. Clinical manifestation

Salmonellosis is a type of food infection that can occur when you eat foods that contain *Salmonella* bacteria. Once ingested, the bacteria may initiate infection and cause illness. The illness's possibility and severity depend largely on the dose, the host's resistance and the specific *Salmonella* strain causing the disease. The bacteria are disseminated by direct contact with the animal or human excrement through faecal to the oral pathway or spread out indirectly by ingesting food contaminated with *Salmonella* from faeces or raw food through cross-contamination. Clinical manifestation in human can be classified into four syndromes: gastroenteritis, enteric fever, septicaemia and asymptomatic chronic carriage.

5.1 Gastroenteritis

Non-typhoidal salmonellosis caused gastroenteritis, a condition commonly called as food poisoning. It is a condition resulted from the inflammation of the gastrointestinal tract spread by faecal to an oral route such as enteric fever. Frequently related serovars reported caused gastroenteritis related to outbreaks are Enteritidis, Typhimurium and Heidelberg [37]. Non-typhoidal salmonellosis is the leading cause of death and hospitalisations among other foodborne pathogens in the USA [1], and the estimated cases of non-typhoidal salmonellosis worldwide massively exceed the enteric fever cases.

The incubation period is 12–72 hours as a result of ingesting tainted food or drinking tainted water and dose-dependent on bacteria that infect the intestines [38]. Symptomatic disease in healthy adults occurs if they are being infected with 10^6–10^8 CFU/mL *Salmonella*. Common symptoms are diarrhoea, vomiting, abdominal pain, nausea, myalgia and headache. In addition, chills and fever within 38–39°C can also occur to the patient. In severe cases, it can lead to severe dehydration and bloody diarrhoea in rare cases. The duration of the symptoms varied from 2 to 7 days but generally resolved by itself without the need for treatment within a week.

5.2 Enteric fever

Enteric fevers are severe systemic forms of salmonellosis and occasionally life-threatening illness. Enteric fever that is caused by *S*. Typhi and *S*. Paratyphi A, B or C infections is called as typhoid fever and paratyphoid fever, respectively [39–41] . Paratyphoid fever is a similar illness causing a milder form of enteric fever compared to the typhoid fever. *S*. Typhi is responsible for causing the most endemic and epidemic cases of enteric fever worldwide with 200,000 deaths and 23 million illness cases per year [42].

The incubation period varied from 6 to 30 days after infection [43] giving rise to symptoms such as gradual fever (38–40°C) over several days, headache, hepatosplenomegaly, myalgias, diarrhoea and constipation when the onset of the systemic disease takes place. Some people develop a transient skin rash with rose-coloured spots, which can be confused with malaria. Therefore, typhoid fever should be suspected in a traveller who is unresponsive to anti-malarial treatment. If left untreated, the symptoms can last for weeks or months. Without treatment, symptoms may last from weeks or months, and it can be deadly [44]. The fatality rate was reported to be at 10–30% if left without treatment but improved to 1–4% fatality in treated patients [45, 46].

5.3 Bacteraemia/septicaemia

Bacteraemia is the presence of viable bacteria in the bloodstream that may occur through a wound, injection or a surgical procedure. Septicaemia is referred to the presence and proliferation of germs in the blood. Septicaemia is a medical term that refers to blood poisoning. *Salmonella* bacteraemia is a condition in which the presence of *Salmonella* bacteraemia in the blood elicits a systematic inflammatory response that can be fatal. It is an intermediate stage of infection in which the patient is not showing any symptom and the bacteria cannot be isolated from faecal specimens. Bacteraemia can further progress to septicaemia whereby the bacteria multiply in the blood and giving symptoms such as chills, fever, high respiration rate or very fast heart rate.

Bacteraemia can be caused by all *Salmonella* subspecies but is more commonly associated with *S*. Choleraesuis, *S*. Paratyphi, *S*. Typhi and *S*. Dublin. The increased risk is seen in old, young and immunocompromised persons. The severity of the infection depends on the bacterial dose, immune response of the patient and the virulence of the *Salmonella* strain [47]. The severe development of septicaemia was reported higher to occur in cancer patients and immunecompromised individual infected with human immunodeficiency virus (HIV) [48, 49].

5.4 Chronic carrier

After treatment for salmonellosis, some patients become *Salmonella* carriers and shed faeces with *Salmonella* for an extended time, making them a reservoir or carrier for the pathogen, thus making them a chronic carrier. *Salmonella* can continue to be excreted in stool for many weeks following resolution of an initial diarrheal episode without symptoms exhibited by the patients. The factors of the host and pathogen that influence the occurrence of carrier state study are limited; hence, the condition is poorly understood [48]. Chronic carrier state occurred higher in patients infected with *S*. Typhi compared to non-Typhi. About 1–4% of patients that recovered from typhoid fever become chronic carriers, while only 0.2–0.6% of patients infected with non-typhoidal salmonellosis progress into the chronic carrier [50]. The chronic carrier state is associated with carcinoma of the gallbladder, which the host could form into the chronic carriage [51].

6. Epidemiology

Epidemiology is the study of the distribution and causative agents of disease and the application of the study to the control of diseases. The epidemiology of salmonellosis cases differs extensively on the type of *Salmonella* spp. implicated. Annually, enteric fever was estimated to cause 200,000 deaths in 22 million illness cases, which mainly occurred in non-developed countries [45] and is low in developed countries. Enteric fever is endemic in many places on the African and Asian continents, as well as in countries throughout Europe, Central and South America and the Middle East. The prevalence and fatality rate caused by the enteric fever may vary greatly from one location to another. Enteric fever is uncommon in the USA and certain European nations, with less than 10 *Salmonella* cases reported per 100,000 people annually. The majority of cases reported in these nations are linked to travel, with foreigners or travellers returning from Pakistan, Africa or India, bringing the disease with them [45].

Contrary, non-typhoidal salmonellosis (NTS) incidence is increasing and continues to lead the gastroenteritis cases worldwide affecting 155,000 deaths from 93.8 million cases estimated every year [37]. Epidemiology data are well documented in developed countries, such as the USA and the countries of Europe, but are poorly compiled in less-developed countries like Asian and African countries. Owing to less effective monitoring systems, statistics on salmonellosis incidence are limited in countries of Asia, Africa and South and Central America, where only 1–10% of cases are reported [52–55]. It was reported that the most frequent serotype in Asia and Africa was *Salmonella* Enteriditis, accounting for 38% and 26% of the clinical isolates, respectively. NTS disease is an extremely serious infection in Vietnam, and the high death rate (26%) is comparable to the incidence in sub-Saharan Africa, which is a significant risk factor for both infection and mortality in HIV-infected individuals [54].

7. Pathogenicity

Pathogenicity of *Salmonella* is dependent on the serovar and the host. However, factors that influencing serovar–host specificity are not well known [56]. The basic of *Salmonella* virulence mechanism is associated with the invasion of intestinal mucosa and multiplication in gut-associated lymphoid tissue (GALT). *Salmonella* will invade non-phagocytic cells in the intestine by promoting their self-uptake in a complicated and dynamic process similar to phagocytosis [57].

Salmonella infection in human can cause either systemic disease with rare association with food poisoning or one that can cause enteritis or localised disease. In oral infection, *Salmonella* must go through a variety of host defence mechanisms and different environments in the stomach during the progression of infection before successfully entering the intestinal tract. *Salmonella* adapted to these settings by using a broad variety of genes that may be reflected as virulence determinants, including *Salmonella*-specific virulence genes, housekeeping genes and regulatory genes. Virulence genes involved in invasion and critical for intracellular survival are grouped in large chromosomal DNA regions termed as *Salmonella* pathogenicity islands (SPIs). SPIs often exist in huge clusters of genes that are found in the vast chromosomal DNA regions that contribute to a certain virulence phenotype that manifests at a given period during infection [58].

Significant pathogenicity islands in *Salmonella* are SPI-1 and SPI-2. SPI-1 being the most well-defined SPI and required for virulence encodes the type III protein secretion system (T3SS), which injects effector proteins into host cells and provides the essential mechanism for intestinal invasion and enteritis formation [59]. The T3SS is the most significant *Salmonella* virulence factor. SPI-1 genes are involved in host cell invasion, immune cell recruitment, apoptosis, and biofilm formation, while other transcription factors encoded outside SPI-1 engage in the expression of SPI-1-encoded genes. SPI-1's regulatory network is intricate and extremely important [37]. SPI-1's ubiquity is conserved and essential for *Salmonella* pathogenicity, as shown by its direct role in invasion. The T3SS of SPI-1 and SPI-2 has been suggested to be inversely regulated [52, 53].

This is an appealing hypothesis as *Salmonella* systemic infection to a host must first infiltrate M cells, where SPI-1 expression is required, and then replicate within macrophages, where SPI-2 expression is required. Mutations in SPI-2 genes encoding the type III secretion apparatus, on the other hand, diminish the expression of genes encoding a transcriptional activator of SPI-1 (*sipC*, *prgK* and *hilA*), suggesting the interaction between SPI-1 and SPI-2 [60]. Over the previous decade, around 30 SPI-2 T3SS effectors have been discovered. Thirteen of them are involved in the regulation of *Salmonella*-containing vacuole (SCV) membrane dynamics, the location of *Salmonella*-containing vacuole (SCV) inside host cells, immunological modulation, cytoskeletal changes and the motility of infected cells, among other things [61].

8. Route of contamination by *Salmonella*

Salmonella is broadly spread in various food types and extensively distributed in the environment. The most common vehicle for *Salmonella* includes poultry, eggs, livestock animal and dairy products [62–64]. The contamination of *Salmonella* can occur at various points along the food chain route, as described in the diagram in **Figure 3**.

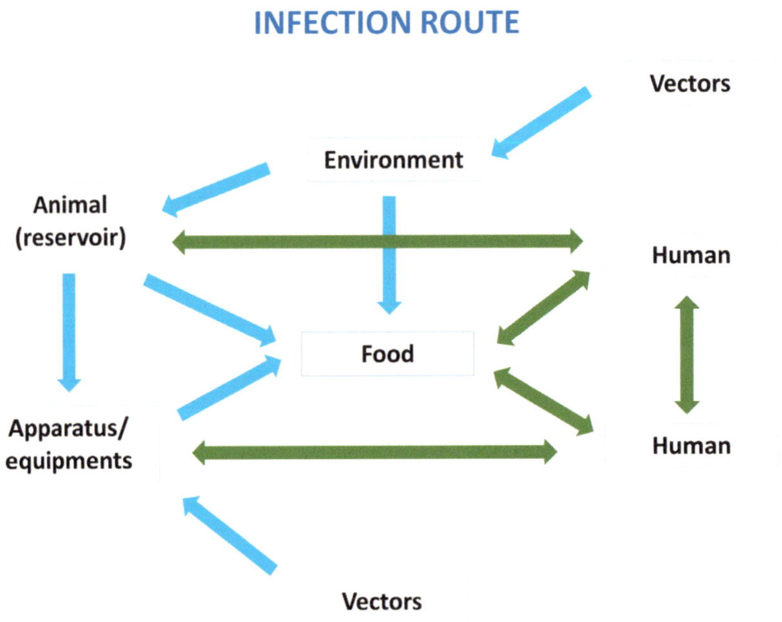

Figure 3.
Various points in food chain where Salmonella *contamination could occur.*

Food that is based on poultry forms the key reservoir of *Salmonella* and poses a risk to be transferred to other medium [65]. The host that the bacteria colonise ranges from wild birds to domestic animals [66, 67]. In most scenarios, *Salmonella* bacteria multiply in chicken and poultry, in which they then become the reservoir for the pathogen (EFSA, 2010). Colonisation in the intestines of the animal becomes the key source of contamination in many points and is typically widespread in the abattoir and poultry processing facilities [68].

9. Conclusion

Salmonella is a gastrointestinal microorganism with numerous abilities to infect and survive in human and animal hosts. This chapter optimistically gives a better insight of *Salmonella's* history background, nomenclature, characteristic, clinical manifestation, epidemiology, pathogenicity and the possible route of contamination. It is worth noting that, despite the advancement in sanitary procedures and quality control in food processing and manufacturing, the infection of these gram-negative bacteria still triggers increase in morbidity and mortality in humans worldwide. Therefore, increased attention and surveillance of this dangerous pathogen should be emphasised and strengthened for the better management of the disease.

Conflict of interest

The authors declare no conflict of interest.

Author details

Mohd Afendy Abdul Talib*, Son Radu, Cheah Yoke Kqueen
and Farinazleen Mohamad Ghazali
Biodiagnostic and Biosensor Programme, Biotechnology and Nanotechnology
Research Centre, MARDI HQ, Persiaran MARDI-UPM, Serdang, Selangor, Malaysia

*Address all correspondence to: fendy@mardi.gov.my

References

[1] Scallan E, Hoekstra RM, Angulo FJ, Tauxe RV, Widdowson MA, Roy SL, et al. Foodborne illness acquired in the United States—Major pathogens. Emerging Infectious Diseases. 2011;**17**(1):7

[2] WHO Estimates of the Global Burden of Foodborne Diseases. Foodborne Disease Burden Epidemiology Reference Group 2007-2015. U.S.: World Health Organization; 2015. Available from: https://apps.who.int/iris/bitstream/handle/10665/199350/9789241565165_eng.pdf

[3] Hu L, Kopecko DJ. Typhoid *Salmonella*. In: Millotis MD, Bier JW, editors. International Handbook of Foodborne Pathogens. New York, USA: Marcel Dekker, Inc; 2003. pp. 151-165

[4] Khan S, Chousalkar KK. *Salmonella* typhimurium infection disrupts but continuous feeding of Bacillus based probiotic restores gut microbiota in infected hens. Journal of Animal Science and Biotechnology. 2020;**11**(1):1-16

[5] Argüello H, Carvajal A, Álvarez-Ordóñez A, Jaramillo-Torres HA, Rubio P. Effect of logistic slaughter on *Salmonella* contamination on pig carcasses. Food Research International. 2014;**55**:77-82

[6] Grassl GA, Valdez Y, Bergstrom KSB, Vallance BA, Finlay BB. Chronic enteric *Salmonella* infection in mice leads to severe and persistent intestinal fibrosis. Gastroenterology. 2008;**134**(3):768-780. E2

[7] Amagliani G, la Guardia ME, Dominici S, Brandi G, Omiccioli E. *Salmonella* Abortusovis: An epidemiologically relevant pathogen. Current Microbiology. 2021;**79**(1):1-7

[8] Gojam A, Tulu D. Infectious causes of abortion and its associated risk factor in sheep and goat in Ethiopia. International Journal of Veterinary Science & Technology. 2020;**4**:007-0012

[9] Glawischnig W, Lazar J, Wallner A, Kornschober C. Cattle-derived *Salmonella* enterica serovar Dublin infections in red foxes (Vulpes vulpes) in Tyrol, Austria. Journal of Wildlife Diseases. 2017;**53**(2):361-363

[10] Kudirkiene E, Sørensen G, Torpdahl M, de Knegt LV, Nielsen LR, Rattenborg E, et al. Epidemiology of *Salmonella enterica* serovar Dublin in cattle and humans in Denmark, 1996 to 2016: A retrospective whole-genome-based study. Applied and Environmental Microbiology. 2020;**86**(3):e01894-19

[11] Vohra P, Vrettou C, Hope JC, Hopkins J, Stevens MP. Nature and consequences of interactions between *Salmonella enterica* serovar Dublin and host cells in cattle. Veterinary Research. 2019;**50**(1):1-11

[12] Eberth CJ. Die Organismen in den Organen bei Typhus abdominalis. Archiv für Pathologische Anatomie und Physiologie und für Klinische Medicin. 1880;**81**(1):58-74

[13] Hardy A. Food, hygiene, and the laboratory. A short history of food poisoning in Britain, circa 1850-1950. Social History of Medicine: The Journal of the Society for the Social History of Medicine. 1999;**12**(2):293-311

[14] Kass EH. A brief perspective on the early history of American infectious disease epidemiology. The Yale Journal of Biology and Medicine. 1987;**60**(4):341

[15] Brands DA, Alcamo IE. Heyman DL. *Salmonella*. Philadelphia, USA: Chelsea House Publishers; 2006. pp. 6-74

[16] Grimont PAD, Grimont F, Bouvet P. In: Wray C, Wray A, editors. Taxonomy of the Genus *Salmonella* in *Salmonella* in Domestic Animal. New York, USA: CABI Publishing; 2000. pp. 1-18

[17] Gal-Mor O, Boyle EC, Grassl GA. Same species, different diseases: How and why typhoidal and non-typhoidal *Salmonella enterica* serovars differ. Frontiers in Microbiology. 2014;5:1-10

[18] Antigenic Formulas of the *Salmonella Serovars*. Available from: https://agris.fao.org/agris-search/search.do?recordID=US201300015921 [Accessed: 23 February 2022]

[19] Popoff MY, Le Minor L. Antigenic Formulas of the *Salmonella* Serovars, 8th Revision. World Health Organization Collaborating Centre for Reference and Research on *Salmonella*, Paris. 2001

[20] Grimont PA, Weill FX. Antigenic formulae of the *Salmonella* serovars. WHO Collaborating Centre for Reference and Research on *Salmonella*. 2007;**9**:1-66

[21] Agbaje M, Begum RH, Oyekunle MA, Ojo OE, Adenubi OT. Evolution of Salmonella nomenclature: A critical note. Folia Microbiologica. 2011;**56**(6):497-503

[22] Brenner FW, Villar RG, Angulo FJ, Tauxe R, Swaminathan B. *Salmonella* nomenclature. Journal of clinical Microbiology. 2000;**38**(7):2465-2467

[23] Deb M, Kapoor L. *Salmonella* nomenclature seen in the literature. Indian Journal of Medical Microbiology. 2005;**23**(3):204-205

[24] Su L-H, Chiu C-H. *Salmonella*: Clinical importance and evolution of nomenclature. Chang Gung Medical Journal. 2007;**30**(3):210-219

[25] Popoff MY, Bockemühl J, Gheesling LL. Supplement 2002 (no. 46) to the Kauffmann-white scheme. Research in Microbiology. 2004;**155**(7):568-570

[26] Lee KM, Runyon M, Herrman TJ, Phillips R, Hsieh J. Review of *Salmonella* detection and identification methods: Aspects of rapid emergency response and food safety. Food Control. 2015;**47**: 264-276

[27] Collee JG, Duguid JP, Fraser AG, Marmion BP, editors. Mackie and McCartney. Practical Medical Microbiology. 13th ed. Chichester: Churchill Livingston; 1989. p. 918

[28] Temitope OO, Boligon AA. Comparative study of antibacterial and phytochemical properties of Nigerian medicinal plants on *Salmonella bongori* and *Salmonella enteritidis* isolated from poultry Feaces in Owo local government. Ondo state, Nigeria. Archives of Current Research International. 2015;**2**(1):1-11

[29] Rosenberg E, DeLong EF, Thompson F, Lory S, Stackebrandt E. The prokaryotes: Prokaryotic biology and symbiotic associations. The Prokaryotes: Prokaryotic Biology and Symbiotic Associations. 2013;**6**:1-607

[30] Uzzau S, Brown DJ, Wallis T, Rubino S, Leori G, Bernard S. Host adapted serotypes of *Salmonella enterica*. Epidemiology & Infection. 2000;**125**(2):229-255

[31] Edwards RA, Olsen GJ, Maloy SR. Comparative genomics of closely related salmonellae. Trends in Microbiology. 2002;**10**(2):94-99

[32] Spector MP, Kenyon WJ. Resistance and survival strategies of

Salmonella enterica to environmental stresses. Food Research International. 2012;**45**(2):455-481

[33] Finn S, Condell O, McClure P, Amézquita A, Fanning S. Mechanisms of survival, responses and sources of *Salmonella* in low-moisture environments. Frontiers in Microbiology. 2013;**4**(331):1-15

[34] Tadapaneni RK, Syamaladevi RM, Villa-Rojas R, Tang J. Design of a novel test cell to study the influence of water activity on the thermal resistance of *Salmonella* in low-moisture foods. Journal of Food Engineering. 2017;**208**:48-56

[35] Mattick KL, Jørgensen F, Wang P, Pound J, Vandeven MH, Ward LR, et al. Effect of challenge temperature and solute type on heat tolerance of *Salmonella* serovars at low water activity. Applied and Environmental Microbiology. 2001;**67**(9):4136

[36] D'Aoust J-Y. Psychrotrophy and foodborne *Salmonella*. International Journal of Food Microbiology. 1991;**13**(3):207-215

[37] Majowicz SE, Musto J, Scallan E, Angulo FJ, Kirk M, O'Brien SJ. The global burden of nontyphoidal *salmonella* gastroenteritis. Clinical Infectious Diseases. 2010;**50**(6):882-889

[38] Besser JM. Salmonella epidemiology: A whirlwind of change. Food Microbiology. 2018;**71**:55-59

[39] Wain J, Hendriksen RS, Mikoleit ML, Keddy KH, Ochiai RL. Typhoid fever. The Lancet. 2015;**385**(9973):1136-1145

[40] Andino A, Hanning I. *Salmonella enterica*: Survival, colonization, and virulence differences among serovars. The Scientific World Journal. 2015;**2015**:1-16

[41] Sanderson KE, Liu SL, Tang L, Johnston RN. *Salmonella* Typhi and *Salmonella* Paratyphi A. Molecular Medical Microbiology: Second Edition. 2014;**2-3**:1275-1306

[42] González-López JJ, Piedra-Carrasco N, Salvador F, Rodríguez V, Sánchez-Montalvá A, Planes AM, et al. ESBL-producing *Salmonella enterica* Serovar Typhi in Traveler returning from Guatemala to Spain. Emerging Infectious Diseases. 2014;**20**(11):1920

[43] Jacobs Slifka KM, Blackstock A, Nguyen V, Schwensohn C, Gieraltowski L, Mahon BE. Estimating the incubation period of *Salmonella* urinary tract infections using foodborne outbreak data. Foodborne Pathogens and Disease. 2020;**17**(10):628-630

[44] Mayer CA, Neilson AA. Typhoid and paratyphoid fever: Prevention in travellers. Australian Family Physician. 2010;**39**(11):847-851

[45] Crump JA, Luby SP, Mintz ED. The global burden of typhoid fever. Bulletin of the World Health Organization. 2004;**82**(5):353

[46] Stanaway JD, Parisi A, Sarkar K, Blacker BF, Reiner RC, Hay SI, et al. The global burden of non-typhoidal *salmonella* invasive disease: A systematic analysis for the global burden of disease study 2017. The Lancet Infectious Diseases. 2019;**19**(12):1312-1324

[47] Kurtz JR, Goggins JA, McLachlan JB. *Salmonella* infection: Interplay between the bacteria and host immune system. Immunology Letters. 2017;**190**:50

[48] Dhanoa A, Fatt QK. Non-typhoidal *Salmonella* bacteraemia: Epidemiology, clinical characteristics and its' association with severe immunosuppression.

Annals of Clinical Microbiology and Antimicrobials. 2009;**8**:1-15

[49] Mori N, Szvalb AD, Adachi JA, Tarrand JJ, Mulanovich VE. Clinical presentation and outcomes of non-typhoidal *Salmonella* infections in patients with cancer. BMC Infectious Diseases. 2021;**21**(1):1-7

[50] Bäumler AJ. The record of horizontal gene transfer. Trends in Microbiology. 1997;**5**(8):318-322

[51] Dutta U, Garg PK, Kumar R, Tandon RK. Typhoid carriers among patients with gallstones are at increased risk for carcinoma of the gallbladder. American Journal of Gastroenterology. 2000;**95**(3):784-787

[52] Boore AL, Hoekstra RM, Iwamoto M, Fields PI, Bishop RD, Swerdlow DL. *Salmonella* enterica infections in the United States and assessment of coefficients of variation: A novel approach to identify epidemiologic characteristics of individual serotypes, 1996-2011. PLoS One. 2015;**10**(12):1-11

[53] Shafini AB, Son R, Mahyudin NA, Rukayadi Y, Zainazor T. Prevalence of *Salmonella* spp. in chicken and beef from retail outlets in Malaysia. International Food Research Journal. 2017;**24**(1):437-449

[54] Phu Huong Lan N, le Thi PT, Nguyen Huu H, Thuy L, Mather AE, Park SE, et al. Invasive non-typhoidal *Salmonella* infections in Asia: Clinical observations, disease outcome and dominant serovars from an infectious disease hospital in Vietnam. PLOS Neglected Tropical Diseases. 2016;**10**(8):e0004857

[55] Crump JA, Heyderman RS. A perspective on invasive *Salmonella* disease in Africa. Clinical Infectious Diseases: An official publication of the Infectious Diseases Society of America. 2015;**61**(Suppl. 4):S240

[56] Barrow PA, Methner U, editors. In: Vaccination against *Salmonella* infections in food animals: Rationale, theoretical basis and practical application in *Salmonella* in Domestic Animals. 2nd ed. New York, USA: CABI Publishing; 2000. pp. 455-475

[57] Hansen-Wester I, Stecher B, Hensel M. Type III secretion of *Salmonella* enterica serovar typhimurium translocated effectors and SseFG. Infection and Immunity. 2002;**70**(3):1403-1409

[58] Hansen-Wester I, Hensel M. *Salmonella* pathogenicity islands encoding type III secretion systems. Microbes and Infection. 2001;**3**(7):549-559

[59] LeBlanc M-A, Fink MR, Perkins TT, Sousa MC. Type III secretion system effector proteins are mechanically labile. Proceedings of the National Academy of Sciences. 2021;**118**(12):e2019566118

[60] Deiwick J, Nikolaus T, Shea JE, Gleeson C, Holden DW, Hensel M. Mutations in *Salmonella* pathogenicity island 2 (SPI2) genes affecting transcription of SPI1 genes and resistance to antimicrobial agents. Journal of Bacteriology. 1998;**180**(18):4775-4780

[61] Figueira R, Holden DW. Functions of the *Salmonella* pathogenicity island 2 (SPI-2) type III secretion system effectors. Microbiology. 2012;**158**(5):1147-1161

[62] Gast RK, Regmi P, Guraya R, Jones DR, Anderson KE, Karcher DM. Contamination of eggs by *Salmonella* Enteritidis in experimentally infected laying hens of four commercial genetic lines in conventional cages and enriched

colony housing. Poultry Science. 2019;**98**(10):5023-5027

[63] Heredia N, García S. Animals as sources of food-borne pathogens: A review. Animal Nutrition. 2018;**4**(3):255

[64] Jones FT. A review of practical *Salmonella* control measures in animal feed. Journal of Applied Poultry Research. 2011;**20**(1):102-113

[65] Park SH, Aydin M, Khatiwara A, Dolan MC, Gilmore DF, Bouldin JL, et al. Current and emerging technologies for rapid detection and characterization of *Salmonella* in poultry and poultry products. Food Microbiology. 2014;**38**: 250-262

[66] Oloya J, Theis M, Doetkott D, Dyer N, Gibbs P, Khaitsa ML. Evaluation of *Salmonella* occurrence in domestic animals and humans in North Dakota (2000-2005) in Foodborne Pathogens and Disease. USA: Mary Ann Liebert, Inc. 2017;**4**(4):551-563

[67] Hughes LA, Shopland S, Wigley P, Bradon H, Leatherbarrow AH, Williams NJ, et al. Characterisation of *Salmonella enterica* serotype typhimurium isolates from wild birds in northern England from 2005-2006. BMC Veterinary Research. 2008;**4**(1):1-10

[68] Rothrock MJ, Ingram KD, Gamble J, Guard J, Cicconi-Hogan KM, Hinton A. The characterization of *Salmonella enterica* serotypes isolated from the scalder tank water of a commercial poultry processing plant: Recovery of a multidrug-resistant Heidelberg strain. Poultry Science. 2015;**94**(3):467-472

Section 3

Enterobacteria

Chapter 6

Phenotypic Characterisation of Carbapenemases Produced by Enterobacteria Isolated from Patients of the Medico-Social Centre of the National Social Insurance Fund of Maroua: Cameroon

Daoudou Bakari, Jojo Mbala Ekamba, Mamoudou Hamadou, Yaya Garga, Martin-Paul Baane and Félicité Obono

Abstract

The aim of this study is to determine the types of carbapenemases moving around the city of Maroua with a view to contribute to the development of a control strategy against the enterobacteria that produce them. The investigation carried out on the biological samples showed that 5.97% of the sample contained carbapenem-resistant microorganisms. This includes 2.20% of urine samples, 0.94% of osteitis samples, 0.63% of wound pus samples, 1.26% of stool samples and 0.94% of blood samples. The microorganisms responsible for this resistance to carbapenems are 5.26% for each of species *Arizona*, *Citrobacter braakii*, *Enterobacter gergoviae*, *P. vulgaris*, and Serratia ficaria, 26.32% for the species *E. gergoviae* and *P. mirabilis* and 21.05% for the species *S. odorifera* 1. All these enterobacteria produce at least one carbapenemase, which 36.84% are of the KPC type, 10.53% of the OXA-48 or OXA-181 type and 52.63% of types that could not be determined by the algorithm proposed by Nordmann et al. used for this purpose. The types of carbapenemases determined in this revealed 11 substrates and inhibition profiles associated with their production. This highlighted the difficulty of applying an inhibition law in situ in the context of probabilistic antibiotic therapy.

Keywords: carbapenemases, enterobacteriaceae, substrate profile

1. Introduction

The increasing complexity of the phenomenon of resistance of enterobacteria against beta-lactamines in Cameroon will be a major public health problem if nothing is done. The study of the enzymatic systems involved in inducing these resistances in the cities of Yaoundé, Ngaoundéré and Douala has shown evolutions in space and time [1–3]. In the city of Maroua, the alarm was raised with the identification of multi-resistant microorganisms in dairy products and raw meat sold in the city [4, 5]. These resistances have been attributed to the production of high-level cephalosporinases and b-lactamases including Carbapenemases [5, 6]. The concern in relation to these observations is that, microorganisms with similar profiles have started to be isolated from biological samples at the National Social Insurance Fund (MSC-NSIF) of Maroua. Moreover, cases of death were noted due to infections by this category of Enterobacteriaceae before the end of the analysis of the samples, although routine antibiotics were administered (MSC-NSIF patient files). The emergence of carbapenemase-producing Enterobacteriaceae (EPC) in the city of Maroua could in the future, become the main factor of therapeutic failures. In a socio-economic context where probably antibiotic therapy remains the most widely used strategy, knowledge of the enzyme systems involved in these enterobacteria could make patient management more effective. The objective of this study is to determine the types of carbapenemases moving around in the city of Maroua with a view to developing a strategy control to fight against the enterobacteria that produce them.

2. Materials and methods

2.1 Study site

The study took place at the MCS-NSIF (Medico-Social Centre of National Social Insurance Fund of Maroua) in Maroua, which is located between 10°35'North latitude and 14°19'East longitude [7]. The MCS-NSIF is one of the hospitals in the distric of Maroua I first. It is adjacent to the road that connects the *'Djarma'* crossroads to the third and northernmost entrance of the SODECOTON company. This hospital receives among its patients, those who clinical examinations require the realisation of the antibiogram.

The requirement of an antibiogram by the clinician was the criteria retained for the choice of the samples to be analysed. In compliance with this requirement, a sample of 318 biological samples was taken from patients received at the MCS-NSIF laboratory in Maroua. This sample consisted of 123 urine samples, 71 vaginal samples, 70 stool samples, 27 blood samples, 13 osteitis pus samples, 9 urethral samples and 5 wound pus samples. The material collection were done between the month of january and febuary in 2018.

2.2 Isolation, purification and selection of resistant carbapenem strains

2.2.1 Isolation of enterobacterial strains

The plating technique carried out near a flame maintained by a Bunsen burner was used to isolate the strains of interest [8]. Biological material that remained attached to the sterile loop handle was streaked onto MacConkey agar in a Petri dish. For the microorganisms to be isolated from the blood, a pre-culture in bovine heart-brain infusion incubated at 37°C for 18 hours in an oven preceded the implementation of the technique.

2.2.2 Purification of enterobacterial strains

The quadrant method was used to purify the strains of interest [8]. One of the colonies from among those having the same appearance during the isolation phase was used for this purpose. The first quadrant was formed by inoculation in tight streaks using a sterile loop. The loop used at this stage of the operation is flamed to red and cooled by touching an unused area of the agar. The Petri dish is rotated at an angle of 90° and the loop is passed once through the first quadrant to form the second quadrant. The same procedure is used to form the other two quadrants.

2.2.3 Selection of strains of interest

The standardised method for determining the susceptibility of bacteria to antibiotics using ertapenem 10 µg, imipenem 10 µg and meropenem 10 µg was used to select the strains of interest [9]. First, a suspension containing 10^6 CFU/mL of bacteria for each of the purified strains was prepared. Swabbing for each of the prepared suspensions was performed on Müller-Hinton (MH) contained in a petri dish. The carbapenem discs were placed at a distance of 3 cm from each other in each of the seeded Petri dishes. All prepared Petri dishes were incubated at 37°C for 18 hours in an oven. It should be noted that the disc quality test was validated on E. coli ATCC 29522 reference strains classified as susceptible. The determination of resistant, intermediate or susceptible traits was based on the comparison between the inhibition diameters obtained and those of the EUCAST reference [9].

2.3 Determination of the enzymatic character of carbapenem resistance

The Carba NP test which is a biochemical colorimetric test was used to demonstrate the enzymatic activity of carbapenem resistance in the strains of interest [10]. A 100 L volume of Tris–HCl B-PER II (Bacterial Protein Extraction Reagent) lysis buffer, 20 mM, pH 7.5 and one colony of bacteria were introduced into each of two 1.5 mL Eppendorf tubes prepared for each strain to be tested. The resulting mixture was homogenised using a 1000 L micropipette. Subsequently, 100 L of solution (A) containing 0.54% (W/V) phenol red and 0.2 mM zinc sulphate was introduced into control tube 1. The same volume of solution (A), this time containing concentrated carbapenem 6 mg/mL, was introduced into test tube 2. Both tubes were incubated at 37°C in the incubator for 2 hours. The appearance of a yellow coloration was interpreted as positive and therefore the presence of carbapenemase, whereas the red coloration was interpreted as negative and therefore the absence of carbapenemase.

2.4 Determination of carbapenemase classes produced by the strains of interest

The classes of carbapenemases produced by the Enterobacteriaceae were determined using phenotypic inhibition and synergy tests [9]. After swabbing on MH, the antibiotics were arranged with a distance of 3 cm between them. These were imipenem (IMP), ertapenem (ETP), meropenem (MRP), amoxicillin (AMX), cefotaxime (CTX), ceftazidime (CAZ), EDTA, clavulanic acid (CMA), cloxacillin (CXC), cefepime (CFP), piperacillin-tazobactam (PIT) and aztreonam (AZT). The elements used in the algorithm to identify the types of carbapenemases produced by the strains of interest were arranged as follows (**Figure 1**):

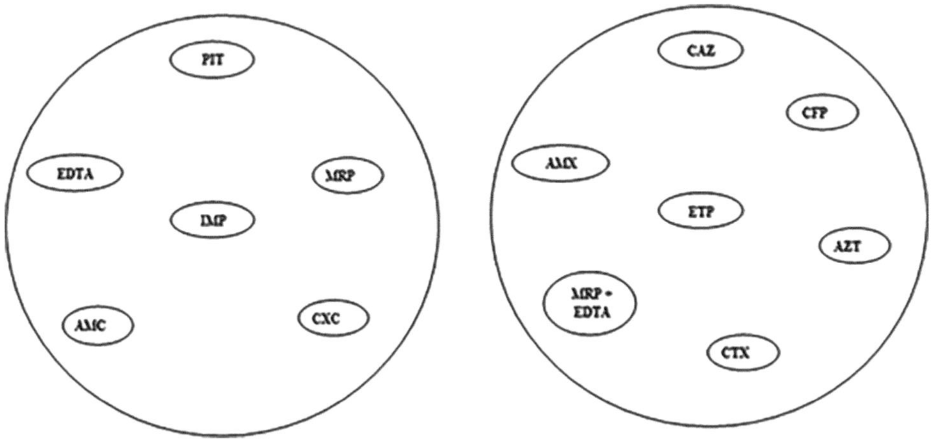

Figure 1.
Layout of the elements of the carbapenemase class identification algorithm.

CLASS	TYPE	AMX	AMC	PIT	CTX	CAZ	IMP	ETP	MRP	AZT
A	KPC	R	S/I	R	R	R	S/I/R	I/R	S/I/R	R
	KPC + BLSE	R	I/R	R	R	R	I/R	I/R	I/R	R
B	IMP/VIM/NDM	R	R	I/R	R	I/R	S/I/R	I/R	S/I/R	S
	IMP/VIM/NDM + BLSE	R	R	I/R	R	R	I/R	R	S/I/R	R
D	OXA-48/OXA-181	R	R	S/I/R	S/I	S	S/I	S/I	S/I	S
	OXA-48/OXA-181 + BLSE	R	R	I/R	R	R	I/R	I/R	I/R	R

Legend: R = Resistant I = Intermediate S = Sensitive [11].

Table 1.
Resistance phenotypes resulting from the expression of carbapenemases reported in Enterobacteriaceae without or with extended-spectrum lactamases.

The classes of carbapenemases produced by the isolated Enterobacteriaceae were determined from the algorithm (**Table 1**).

2.5 Determination of minimum inhibitory concentrations of carbapenems

The E-test, an agar diffusion technique, was used to determine the minimum inhibitory concentrations (MICs) [12]. The commercially available strip was adapted with easily accessible blotting paper. This blotting paper, cut to the size of 10 cm x 1 cm, was divided into 10 zones of equal size (widthwise) by lines obtained with a pencil. This paper was sterilised in an autoclave at 125°C for 15 minutes. To determine the MICs of the ETP, 640 g of antibiotic was introduced into a sterile 10 mL volumetric flask. This mass was dissolved in 5 mL of sterile distilled water measured with a pipette. After complete dissolution, the volume was made up to the mark to obtain a concentrated solution C1 64 mg/L. Solution C1/2 was obtained by removing 1 mL of solution C1 and adding it to a tube containing the same volume of sterile distilled water. Concentrated solution C1/2n was obtained by pipetting 1 mL of the prepared concentrated solution C1/n into 1 mL of distilled water. Once the 10 dilutions had been obtained, 25 L was taken from each tube using a 50 L micropipette and arranged along with the blotting paper in the corresponding zones respecting the gradient (C1, C1/2, C1/4, C1/8, C1/16, C1/32, C1/64, C1/128, C1/256, C1/512). The same procedure was adopted for the determination of MICs for MRP and IMP.

2.6 Identification of carbapenemase-producing strains

The identification of carbapenemase-producing strains of Enterobacteriaceae has followed a three-stage procedure [13].

2.6.1 Orientation of the diagnosis by observing particularly discriminating features of the Enterobacteriaceae

Characteristics such as pigmentation and mucoid of colonies, invasion of solid media by colonies, appearance of small colonies were observed directly on the culture medium after incubation at 37°C for 24 h. For mobility, 20 L of a suspension from a colony dissolved in 1 mL of peptone water and incubated at 37°C for 30 minutes was placed on a slide using a 50 L micropipette and covered with a coverslip. This mount was viewed under the 40X objective of the microscope to observe the movement of the bacteria. Suspicion of enterobacteria was made when the bacteria were either immobile or showed peritrich-like mobility.

For Gram staining, a colony from the culture medium was placed in a thin layer on the slide using a platinum loop. The layer formed was fixed by the flame maintained by the Bunsen burner. This layer was covered with gentian violet for 45 seconds, rinsed with water and then covered again with Lugol's. This Lugol's was cleaned after 45 seconds with 95° alcohol in a wash bottle and then rinsed with water. The washed slide was then covered with Fuchsin for 45 seconds, rinsed again with water, dried and read under a 100X microscope objective. The presence of an enterobacterium was confirmed if a Gram-negative bacillus was observed with bipolar staining.

2.6.2 Revelation of the biochemical characteristics that characterise their metabolism

The biochemical characteristics of the metabolism of the enterobacteria retained after diagnostic orientation were obtained using the API 20E gallery. Firstly, the tubes were moistened by introducing 10 mL of sterile distilled water. A bacterial suspension for each species was prepared by diluting the colonies from MH in 5 mL of sterile distilled water. Each tube in the gallery was inoculated with the corresponding suspension using a sterile Pasteur pipette. They were filled by pressing the Pasteur pipette inwards and to the side to avoid bubbles. The wells for citrate (CIT), Voges Prauskauer (VP), gelatinase (GEL) traits were filled completely (tube and cup) for aerobic conditions. For the Arginine dehydrogenase (ADH), Lysine decarboxylase (LDC), Ornitine decarboxylase (ODC), Hydrogen sulphide (H2S) and Urease (URE) wells, the filling was done only at the level of the tube and the well was filled with paraffin oil to create anaerobic conditions. The whole set was incubated at 37°C in the incubator for 22 hours and then a drop of developer was introduced in some wells. These were FeCl3 in the Tryptophan deaminase (TDA) well, Kovacs reagent in the Indole (IND) well, −naphthol and NaOH in the VP well and Nit1, Nit 2 in the BNit well. The staining obtained in each well provided guidance on the positivity or negativity of the reaction.

2.6.3 Identification

The result of the reactions obtained in each well is fed into the Enterobacteriaceae identification software which displays the species of Enterobacteriaceae responsible for the biochemical properties obtained in the API 20 E gallery wells.

2.7 Data analysis

The data obtained were analysed using SPSS 20 and API 20 E Enterobacteriaceae identification software. The SPSS 20 software was used to convert the experimental results into percentages. This software was also used to calculate Pearson's correlation values between inhibition diameters and carbapenem MICs. The second software was used to determine the species of enterobacteria from the results obtained from the API 20 E gallery.

3. Results and discussion

3.1 Results

3.1.1 Identification of microorganisms and mechanism of resistance to carbapenems

The proportion of biological samples containing carbapenem-resistant microorganisms was 5.97%. This proportion is distributed between urine samples, which represent 2.20%, osteitis pus 0.94%, wound pus 0.63%, stool 1.26% and blood 0.94%. The presence of carbapenem-resistant microorganisms was not observed in urethral and vaginal swabs. The percentage of samples that did not contain carbapenem-resistant microorganisms was 94.03%. This frequency was distributed among urine samples 36.48%, osteitis pus 0.63%, wound pus 3.46%, urethral 2.83%, vaginal 22.33%, stool 20.75% and blood 7.55% (**Table 2**).

The species of enterobacteria responsible for carbapenem resistance in biological samples are variously distributed. Urine samples contain 36.84% of carbapenem-resistant microorganisms. This percentage is distributed between the species

Biological samples	Frequency (%)		Totals
	Containing resistant carbapenem enterobacteria	Not containing resistant carbapenem enterobacteria	
Urine	2.20	36.48	38.68
Pus from osteitis	0.94	0.63	1.57
Pus from wounds	0.63	3.46	4.09
Urethra	0.00	2.83	2.83
Vaginal	0.00	22.33	22.33
Stool	1.26	20.75	22.01
Blood	0.94	7.55	8.49
Totals	**5.97**	**94.03**	**100.00**

Table 2.
Frequency of biological samples with and without carbapenem-resistant Enterobacteriaceae.

Enterobacter gergoviae 10.53%, Enterobacter asburiae 5.26%, Proteus mirabilis 5.26%, Proteus vulgaris 5.26%, Serratia ficaria 5.26% and Serratia odorifera 1 5.26%. The proportion of 21.05% of carbapenem-resistant microorganisms was obtained in osteitis pus. This proportion is represented by the microorganisms *Arizona* 5.26%, *P. mirabilis* 10.53% and *S. odorifera* 1 21.05%. Carbapenem-resistant microorganisms identified in wound pus samples account for 10.53%. These were *E. gergoviae* 5.26% and *P. mirabilis* 5.26%. In blood samples, the proportion of microorganisms of interest is 10.53%. It is represented by *E. gergoviae* 5.26% and *S. odorifera* 1 5.26%. Carbapenem-resistant microorganisms in stool samples represent a proportion of 21.05%. These are Citrobacter braakii 5.26%, *E. gergoviae* 5.26%, *P. mirabilis* 5.26% and *S. odorifera* 1.5.26% (**Table 3**).

3.1.2 Phenotypes of identified carbapenemases

3.1.2.1 Dissemination of identified carbapenemases among enterobacteria

The carbapenemases circulating in the city of Maroua are of several types and in different proportions. The KPC type which represents 36.84% of identified carbapenemases is produced at 5.26% by each of the species *Arizona*, *E. asburiae, P. mirabilis, P. vulgaris, S. ficaria* and at 10.53% by the species *E. gergoviae*. Type OXA-48 or OXA 181 represents 10.53% of all these carbapenemases. The species *P. mirabilis* and *S. odorifera 1* each contributes 5.26% of the production. Eight types of carbapenemases produced by enterobacteria do not fit into the reference algorithm. These are non-determined types (NDPs) 1, 2, 3, 4, 5, 6, 7 and 8. NDE 1 represents 10.53% of all carbapenemases and is produced by *E. gergoviae* and *S. odorifera 1*, which each contributes 5.26%. TND 2, which accounts for 10.53%, is produced by the microorganism *P. mirabilis*. TND 3 produced by a single microorganism, *S. odorifera* 1, is present at 5.26%. TND 4, produced by the microorganism *P. mirabilis* only, occupies

Biological samples	**Enterobacteriaceae species**								
	Arizona	***Citrobacter braakii***	***Enterobacter gergoviae***	***Enterobacter asburiae***	***P. vulgaris***	***P. mirabilis***	***Serratia ficaria***	***S. odorifera 1***	**Totals**
Urine	0.00	0.00	10.53	5.26	5.26	5.26	5.26	5.26	**36.84**
Pus from osteitis	5.26	0.00	0.00	0.00	0.00	10.53	0.00	5.26	**21.05**
Pus from wounds	0.00	0.00	5.26	0.00	0.00	5.26	0.00	0.00	**10.53**
Urethra	0.00	0.00	0.00	0.00	0.00	0.00	0.00	0.00	**0.00**
Vaginal	0.00	0.00	0.00	0.00	0.00	0.00	0.00	0.00	**0.00**
Blood	0.00	0.00	5.26	0.00	0.00	0.00	0.00	5.26	**10.53**
Stool	0.00	5.26	5.26	0.00	0.00	5.26	0.00	5.26	**21.05**
Totals	**5.26**	**5.26**	**26.32**	**5.26**	**5.26**	**26.32**	**5.26**	**21.05**	**100.00**

Table 3.
Frequency of Enterobacteriaceae species identified in the biological samples taken.

Enterobacteria species	Carbapenemase types										Totals
	KPC	OXA-48 ou OXA 181	Undetermined 1	Undetermined 2	Undetermined 3	Undetermined 4	Undetermined 5	Undetermined 6	Undetermined 7	Undetermined 8	
Arizona	5.26	0.00	0.00	0.00	0.00	0.00	0.00	0.00	0.00	0.00	**5.26**
Citrobacter braakii	0.00	0.00	0.00	0.00	0.00	0.00	0.00	5.26	0.00	0.00	**5.26**
Enterobacter gergoviae	10.53	0.00	5.26	0.00	0.00	0.00	0.00	0.00	5.26	5.26	**26.32**
Enterobacter asburiae	5.26	0.00	0.00	0.00	0.00	0.00	0.00	0.00	0.00	0.00	**5.26**
P. mirabilis	5.26	5.26	0.00	10.53	0.00	5.26	0.00	0.00	0.00	0.00	**26.32**
P. vulgaris	5.26	0.00	0.00	0.00	0.00	0.00	0.00	0.00	0.00	0.00	**5.26**
Serratia ficaria	5.26	0.00	0.00	0.00	0.00	0.00	0.00	0.00	0.00	0.00	**5.26**
S. odorifera 1	0.00	5.26	5.26	0.00	5.26	0.00	5.26	0.00	0.00	0.00	**21.05**
Totals	**36.84**	**10.53**	**10.53**	**10.53**	**5.26**	**5.26**	**5.26**	**5.26**	**5.26**	**5.26**	**100.00**

Table 4.
Dissemination of circulating carbapenemase types between the identified enterobacteria species.

5.26%. The TND 5, 6, 7, 8 each represents 5.26% produced respectively by the microorganisms *S. odorifera* 1, *C. braakii*, and *E. gergoviae* for the last two types (**Table 4**).

3.1.2.2 Distribution of carbapenemase types in biological samples

The carbapenemases circulating in the city of Maroua are differently distributed in biological samples. The KPC type was found in 21.05% of urine samples, 5.26% of osteitis pus samples and 10.53% of wound pus samples. For type OXA-48 or OXA-181, 5.26% is present in osteitis pus and 5.26% in blood samples. TND 1 is only found in urine samples at a proportion of 10.53%. TND 2 was present in 5.26% of urine samples and in the same proportion of osteitis pus samples. TND 3 was present in 5.26% of the osteitis pus samples only. TND 4, 5, 6, 7 are only found in stool samples and represent 5.26% each. TND 8 is only found in blood samples and represents 5.26% (**Table 5**).

3.1.3 Substrate and inhibitor profiles

The results of the Carba NP test showed that all the Enterobacteriaceae identified in the biological samples use an enzymatic mechanism as a means of resistance to carbapenems. On the other hand, the study of the substrate and inhibitor profiles highlighted three cases, namely enzymatic activity implying resistance (R), decreased enzymatic activity leading to intermediate resistance (I) and a complete absence of enzyme activity implying sensitivity (S).

3.1.3.1 Carbapenemase substrate and inhibitor profiles

The carbapenemase KPC has described two different profiles defined as (P1 and P2). The P1 profile is observed with the microorganisms *Arizona* isolated from osteitis pus

	Carbapenemase types										
Biological samples	**KPC**	**OXA-48 or OXA 181**	**Undetermined 1**	**Undetermined 2**	**Undetermined 3**	**Undetermined 4**	**Undetermined 5**	**Undetermined 6**	**Undetermined 7**	**Undetermined 8**	**Totals**
Urine	21.05	0.00	10.53	5.26	0.00	0.00	0.00	0.00	0.00	0.00	36.84
Pus from osteitis	5.26	5.26	0.00	5.26	5.26	0.00	0.00	0.00	0.00	0.00	21.05
Pus from wounds	10.53	0.00	0.00	0.00	0.00	0.00	0.00	0.00	0.00	0.00	10.53
Blood	0.00	5.26	0.00	0.00	0.00	0.00	0.00	0.00	0.00	5.26	10.53
Stool	0.00	0.00	0.00	0.00	0.00	5.26	5.26	5.26	5.26	0.00	21.05
Totals	36.84	10.53	10.53	10.53	5.26	5.26	5.26	5.26	5.26	5.26	100.00

Table 5.
Distribution of carbapenemases in biological samples.

samples, *P. mirabilis* isolated from wound pus samples, *S. ficaria, E. gergoviae, E. asburiae, P. vulgaris* isolated from urine samples. The P1 profile, which is characterised by enzymatic activity on all substrates (ETP, IMP, MRP, AMX, CAZ, CTX, CFP) is not inhibited by the carbapenemase inhibitors used (AMC, CXC, AZT, EDTA, PIT). The MICs in the P1 profile are greater than 64 mg/L for ETP, 16 mg/L for IMP and 64 mg/L for MRP.

The second P2 profile was observed with *E. gergoviae* isolated from wound pus samples. Enzymatic activity towards all substrates was maintained in the presence of the inhibitors used, except for AMC, for which it was rather reduced. The MIC values here are above 64 mg/L for ETP and 16 mg/L for IMP and MRP respectively.

The identified OXA carbapenemases describe a single substrate and inhibition profile. This P3 profile is observed with *P. mirabilis* and *S. odorifera* 1 isolated from osteitis pus and blood samples respectively. The activity of the enzyme in this profile is observed on certain substrates (AMX, CFP) and in the presence of inhibitors (CXC, AMC, PIT). It is decreased on the substrates (ETP, IMP, MRP) as well as on one of the inhibitors, EDTA. Another characteristic of this profile is the absence of enzymatic activity towards the substrates CAZ, CTX and one of the inhibitors, AZT. The MICs for these microorganisms are greater than 64 mg/L for ETP, 4 mg/L for IMP, 4 mg/L and 8 mg/L for MRP in *P. mirabilis* and *S. odorifera* 1, respectively.

The P4 profile characterising the TND 1 carbapenemase was identified in *E. gergoviae* and *S. odorifera* 1, all isolated from urine samples. The activity of the enzyme is observed towards the substrates ETP, MRP, AMX, CAZ, CFP and in the presence of the inhibitors CXC, AMC, AZT, PIT. This enzymatic activity is diminished in the presence of IMP, one of the inhibitors, EDTA and is absent in the presence of the substrate CTX. The MICs are greater than 64 mg/L for ETP in both microorganisms, equal to 16 mg/L in *E. gergoviae* for IMP and MRP respectively, and 4 mg/L for IMP and 32 mg/L for MRP in *S. odorifera* 1.

Only *P. mirabilis* isolated from urine samples and osteitis pus expressed the P5 profile. This P5 profile associated with the production of the TND 2 carbapenemase shows an absence of enzymatic activity on the CAZ substrate. However, this activity is observed with regard to the substrates (ETP, IMP, MRP, AMX, CTX, CFP) and in the

Enzymes	Espèces d'entérobactéries	Biological samples	Differentiation	ETP	IMP	MRP	AMX	CAZ	CTX	CXC	AMC	AZT	MRP + EDTA	CFP	PIT	Profiles
KPC + BLSE	*Arizona*	Pus from osteitis	Effect on substrate	R	R	R	R	R	R	R	R	R	R	R	R	P1
			DIC (mm)	10	0	13	0	0	18	0	10	12	13	7	12	
			MIC (mg/L)	>64	32	64	/	/	/	/	/	/	/	/	/	
	Enterobacter asburiae	Urine	Effect on substrate	R	R	R	R	R	R	R	R	R	R	R	R	
			DIC (mm)	10	11	13	0	15	0	0	12	0	13	0	11	
			MIC (mg/L)	>64	64	32	/	/	/	/	/	/	/	/	/	
	Enterobacter gergoviae	Urine	Effect on substrate	R	R	R	R	R	R	R	R	R	R	R	R	
			DIC (mm)	10	16	13	8	0	16	0	13	14	13	10	11	
			MIC (mg/L)	>64	16	32	/	/	/	/	/	/	/	/	/	
	Proteus mirabilis	Wound pus	Effect on substrate	R	R	R	R	R	R	R	R	R	R	R	R	
			DIC (mm)	11	13	15	7	16	20	0	15	18	16	14	12	
			MIC (mg/L)	>64	16	64	/	/	/	/	/	/	/	/	/	
	Proteus vulgaris	Urine	Effect on substrate	R	R	R	R	R	R	R	R	R	R	R	R	
			DIC (mm)	8	11	12	0	0	14	0	14	0	13	0	10	
			MIC (mg/L)	>64	>64	64	/	/	/	/	/	/	/	/	/	
	Serratia ficaria	Urine	Effect on substrate	R	R	R	R	R	R	R	R	R	R	R	R	
			DIC (mm)	8	13	12	0	0	13	0	10	12	12	10	16	
			MIC (mg/L)	>64	64	32	/	/	/	/	/	/	/	/	/	
KPC or KPC + BLSE	*E. gergoviae*	Wound pus	Effect on substrate	R	R	R	R	R	R	R	I	R	R	R	R	P2
			DIC (mm)	6	13	15	10	15	15	0	16	0	15	10	12	
			MIC (mg/L)	>64	16	16	/	/	/	/	/	/	/	/	/	

Enzymes	Espèces d'entérobactéries	Biological samples	Differentiation	ETP	IMP	MRP	AMX	CAZ	CTX	CXC	AMC	AZT	MRP + EDTA	CFP	PIT	Profiles
OXA-48 or OXA-181	*P. mirabilis*	Pus from osteitis	Effect on substrate	I	I	I	R	S	S	R	R	S	I	R	R	P3
			DIC (mm)	12	20	18	15	23	27	0	18	27	20	20	13	
			MIC (mg/L)	>64	4	4	/	/	/	/						
	Serratia odorifera 1	Blood	Effect on substrate	I	I	I	R	S	S	R	R	S	I	R	R	
			DIC (mm)	12	20	20	0	23	23	0	12	29	20	22	9	
			MIC (mg/L)	>64	4	8	/	/	/	/	/	/	/	/	/	
Undetermined 1	*E. gergoviae*	Urine	Effect on substrate	R	I	R	R	R	S	R	R	R	I	R	R	P4
			DIC (mm)	16	14	13	0	15	20	0	10	15	16	12	15	
			MIC (mg/L)	>64	16	16	/	/	/	/	/	/	/	/	/	
	S. odorifera 1	Urine	Effect on substrate	R	I	R	R	R	S	R	R	R	I	R	R	
			DIC (mm)	12	21	16	7	15	20	0	18	20	17	19	11	
			MIC (mg/L)	>64	4	32	/	/	/	/	/	/	/	/	/	
Undetermined 2	*P. mirabilis*	Urine	Effect on substrate	R	R	R	R	S	R	R	R	R	R	R	R	P5
			DIC (mm)	6	12	12	0	20	7	0	12	0	12	0	12	
			MIC (mg/L)	>64	32	>64	/	/	/	/	/	/	/	/	/	
		Pus from osteitis	Effect on substrate	R	R	R	R	S	R	R	R	R	R	R	R	
			DIC (mm)	6	11	14	0	20	16	0	12	7	14	11	12	
			MIC (mg/L)	>64	32	32	/	/	/	/	/	/	/	/	/	
Undetermined 3	*S. odorifera 1*	Pus from osteitis	Effect on substrate	R	I	I	R	R	S	R	R	I	I	R	R	P6
			DIC (mm)	12	20	18	0	12	22	0	16	21	19	18	10	
			MIC (mg/L)	>64	4	4	/	/	/	/	/	/	/	/	/	

Enzymes	Espèces d'entérobactéries	Biological samples	Differentiation	ETP	IMP	MRP	AMX	CAZ	CTX	CXC	AMC	AZT	MRP + EDTA	CFP	PIT	Profiles
Undetermined 4	*P. mirabilis*	Stools	Effect on substrate	R	I	I	R	S	S	R	S	R	I	I	R	P7
			DIC (mm)	11	20	20	10	25	21	0	19	20	20	25	12	
			MIC (mg/L)	>64	4	4	/	/	/	/	/	/	/	/	/	
Undetermined 5	*S. odorifera 1*	Stools	Effect on substrate	R	S	I	R	R	S	R	S	I	I	I	R	P8
			DIC (mm)	12	25	20	11	16	26	0	19	22	21	21	11	
			MIC (mg/L)	>64	0,125	2	/	/	/	/	/	/	/	/	/	
Undetermined 6	*Citrobacter braakii*	Stools	Effect on substrate	R	I	S	R	S	S	R	S	S	I	I	R	P9
			DIC (mm)	13	25	22	10	21	28	0	19	27	22	22	12	
			MIC (mg/L)	>64	0,5	4	/	/	/	/	/	/	/	/	/	
Undetermined 7	*E. gergoviae*	Stools	Effect on substrate	R	S	I	R	I	S	R	S	S	I	I	R	P10
			DIC (mm)	12	23	21	12	17	26	0	20	28	21	21	8	
			MIC (mg/L)	>64	0,5	4	/	/	/	/	/	/	/	/	/	
Undetermined 8	*E. gergoviae*	Blood	Effect on substrate	R	R	S	R	S	S	R	R	R	S	I	I	P11
			DIC (mm)	13	16	22	0	22	27	0	16	17	25	22	19	
			MIC (mg/L)	>64	16	0,5	/	/	/	/	/	/	/	/	/	

Table 6.
Substrate and inhibition profiles associated with the production of carbapenemases circulating in Maroua.

presence of all inhibitors (AMC, CXC, AZT, EDTA, PIT). MICs are greater than 64 mg/L for ETP, equal to 32 mg/L for IMP and greater than 64 mg/L for MRP when isolated from urine and equal to 32 mg/L for MRP when isolated from osteitis pus.

The P6 profile is expressed by the *S. odorifera* 1 microorganism isolated from osteitis pus which produces TND 3 carbapenemase. The enzymatic activity of this profile is observed on the substrates ETP, AMX, CAZ and in the presence of the inhibitors CXC, AMC, PIT. This decreased enzymatic activity on the substrates IMP, MRP and in the presence of the inhibitors AZT and EDTA are absent with respect to CTX. The MIC values here are above 64 mg/L for ETP and 4 mg/L for IMP and MRP.

The *P. mirabilis* microorganism isolated from stool samples and producing the TND 4 carbapenemase expresses the P7 profile. This profile is characterised by an enzymatic activity towards each of the two substrates ETP, AMX and towards three inhibitors CXC, AZT, PIT. This enzymatic activity is decreased on the substrates IMP, MRP, CFP and in the presence of the inhibitor EDTA. Finally, no enzymatic activity was observed on the substrates CAZ, CTX and in the presence of the inhibitor AMC. The MIC values for this microorganism are greater than 64 mg/L for ETP and equal to 4 mg/L for IMP and MRP respectively.

The P8 profile described by TND 5 carbapenemase is observed with *S. odorifera* 1 isolated from stool samples. It is characterised by an enzymatic activity towards the substrates ETP, AMX, CAZ and in the presence of the inhibitors CXC, PIT. This enzymatic activity, which is diminished in the presence of the substrates MRP, CFP and the inhibitors AZT, EDTA, is absent on two substrates IMP, CTX and on an inhibitor AMC. The MICs here are greater than 64 mg/L for ETP, equal to 0.125 mg/L for IMP and 2 mg/L for MRP.

The *C. braakii* species producing the TND 6 carbapenemase isolated from stool samples express the P9 profile. The enzymatic activity here is observed on ETP and AMX substrates and in the presence of the inhibitors CXC and PIT. This enzymatic activity is decreased on both substrates MRP, CFP and in the presence of the inhibitor EDTA. It is absent on the substrates IMP, CAZ, CTX and in the presence of the inhibitors AMC, AZT. The MICs here are greater than 64 mg/L for ETP, equal to 0.5 mg/L for IMP and 4 mg/L for MRP.

The P10 profile is observed with the TND 7 carbapenemase produced by *E. gergoviae* isolated from stool samples and is characterised by enzymatic activity on the substrates ETP, AMX and in the presence of the inhibitors CXC, PIT. This enzymatic activity is decreased in the presence of two substrates MRP, CAZ and the inhibitor EDTA. There is no enzymatic activity on two substrates IMP, CTX and on two inhibitors AMC, AZT. The MICs are above 64 mg/L for ETP, 0.5 mg/L for ETP, and 4 mg/L for MRP.

Finally, the P11 profile is always found in *E. gergoviae* isolated from blood samples but which produces the TND 8 carbapenemase. It is characterised by enzymatic activity on three substrates ETP, IMP, AMX and on three inhibitors CXC, AMC, AZT. This enzymatic activity is decreased in the presence of the substrates MRP, CFP and in the presence of the inhibitor PIT. On the other hand, it is absent on the substrates CAZ, CTX and in the presence of the inhibitor EDTA. The MICs are greater than 64 mg/L for ETP, 16 mg/L for IMP and 0.5 mg/L for MRP (**Table 6**).

3.2 Discussion

Biological samples containing carbapenem-resistant Enterobacteriaceae represented 5.97%. This percentage is distributed between urine samples (2.20%), osteitis pus (0.94%), wound pus (0.63%), blood (0.94%) and stool (1.26%). Carbapenem-resistant Enterobacteriaceae were not identified in urethral and vaginal swabs. The high proportion of carbapenem-resistant microorganisms in urine could be explained by the fact that this medium is potentially an extra-digestive reservoir for ESBL-producing Enterobacteriaceae [14]. The emergence of carbapenem resistance in some of the biological samples taken reflects the increasing complexity of the phenomenon in enterobacteria [15]. This complexification of the resistance phenomenon in the city of Maroua had already been observed in bacteria contaminating the food sold there [4, 5]. Several explanations can be found for the emergence of carbapenem resistance in the city of Maroua. The emergence of carbapenem resistance could be the consequence of exponential and uncontrolled use of antibiotics [6, 16–18]. The flow of populations between risk areas (Europe, Asia) and the city of Maroua could also contribute to the importation of strains expressing these types of resistance [19]. The opening of the University of Maroua, which contributes enormously to the migration of populations from various origins to the city, is also a major risk factor for the transport of multidrug-resistant strains of bacteria. The emergence of this type of resistance may finally be due to an exchange of the genes responsible for their expression between bacterial species from the digestive tract or the environment [20]. This exchange can take place via the phenomena of transduction [21], conjugation [22], or transformation [23].

Using API 20 E galleries, *Arizona*, *C. braakii*, *E. gergoviae*, *E. asburiae*, *P. mirabilis*, *P. vulgaris*, *S. ficaria* and *S. odorifera* 1 were identified as the carbapenem-resistant Enterobacteriaceae in the specimens. These Enterobacteriaceae are variously distributed in the samples. The species *E. gergoviae*, *E. asburiae*, *P. mirabilis*, *P. vulgaris*, *S. ficaria*, *S. odorifera* 1 were identified in urine specimens. Those found in osteitis pus were *Arizona*, *P. mirabilis*, *S. odorifera* 1. Two microorganisms, *E. gergoviae and P. mirabilis* were isolated from wound pus samples. The microorganisms isolated from blood were *E. gergoviae*, *S. odorifera 1. Finally, C. braakii*, *E. gergoviae*, *P. mirabilis* and *S. odorifera* 1 were identified in stool samples. The proportions of carbapenem-resistant Enterobacteriaceae in biological samples were 36.84% in urine samples, 21.05% in osteitis pus and stools respectively, and 10.53% in blood and wound pus samples. This distribution in biological samples shows that Enterobacteriaceae are likely to cause deleterious effects in the organism from a variety of environments [24]. The diversity of environments where these enterobacteria have been identified can be explained by the great power of adaptation that characterises them [25] and the multi-resistance to antibiotics that does not facilitate their elimination [16, 17].

The enzymatic mechanism of resistance to carbapenems was demonstrated in 100% of the Enterobacteriaceae that were identified. This observation is in agreement with the fact that enzymatic inactivation of carbapenems is the main mechanism used by enterobacteria to resist their bactericidal effects [26]. The yellow colour change of phenol red used as a colour indicator to show the presence of enzymatic activity on carbapenems has been interpreted as the result of acidification of the reaction medium [27, 28]. This acidification of the reaction medium is a consequence of hydrolysis of the -lactam ring at the amide bond which produces a carboxyl function [29]. The level of expression of this reaction confers certain characteristics to enterobacteria. These characteristics were assessed indirectly on culture media using the inhibition diameters-MIC relationship [9]. The inhibition diameters-MIC correlation for selected carbapenems ($r = 0.578$, $p < 0.01$ for IMP and $r = 0.858$, $p < 0.01$ for MRP) allowed three characteristics to be defined. The first characteristic is resistance to carbapenem, which indicates the presence of enzymatic activity (R). The second characteristic is intermediate resistance which is the result of decreased enzyme activity (I). The third characteristic, marked by an absence of enzyme activity (S), defines the susceptibility of the enterobacteria to carbapenem [9].

The interpretation of the characteristics expressed by the enterobacteria in the presence of the substrates and inhibitors defined by the algorithm used made it possible to highlight three types of carbapenemases in these enterobacteria isolated from biological samples. These are carbapenemases of the KPC, OXA-48 or OXA-181 type and TNDs. The dominant proportion of KPC carbapenemases (36.84%) can be explained by the fact that they are the most abundant and widespread among enterobacteria [30]. They are also characterised by the existence of several variants that differ only by the substitution of one or two amino acids [31]. In contrast, the low percentage of OXA-41 or OXA-181 carbapenemases (10.53%) in the samples can be justified by the fact that this is an enzyme produced from a single auto transferable plasmid that does not carry additional resistance genes [32]. The low proportion of each of the TNDs can be explained by the fact that they are new phenotypes of point synthesis due to the presence of integrons. Integrons sometimes contain transposons from which some transposase-containing Enterobacteriaceae can be naturally genetically engineered to form highly expressed resistance operons [33].

The types of carbapenemases identified are differently distributed in biological samples and between enterobacteria. This random distribution within species of Enterobacteriaceae could be justified by the ease with which resistance-conferring genes diffuse between microorganisms [11]. It is this random distribution that may explain the difficulty in effectively applying probabilistic and/or therapeutic antibiotic therapy in cases of infection with resistant carbapenem enterobacteria [34]. The enzymatic activity of carbapenemases, which is manifested by hydrolysis at the amide bond of the said ring, has made it possible to describe 11 different substrates and inhibition profiles.

The first substrate and inhibition profile, P1, is characterised by enzymatic activity on all carbapenems including monobactam (AZT) used. The fact that this enzymatic activity is not influenced by the presence of EDTA proves that the enzyme does not need a heavy metal to hydrolyse the substrates. These characteristics are unique to KPC-type class A carbapenemases produced from plasmids [35]. It was also observed that the activity of this enzyme is maintained in the presence of its inhibitors PIT and AMC. This observation highlights a synergy of action between the carbapenemase KPC and an ESBL. Indeed, in the presence of a "suicide" inhibitor that serves as a decoy, such as clavunate or tazobactam, the bacteria compensate for the enzymatic deficit by amplifying the synthesis of ESBLs [6, 36]. This hyperproduction can be mediated by mutations in the promoter of the gene and/or by an increase in the number of plasmids carrying the bla gene. These ESBLs would therefore play the known role of multiplying the targets of antibiotics to limit their effectiveness [37]. From the above, it appears that bacteria of the P1 profile have the capacity to produce both KPC-type carbapenemases and ESBLs, all of which are class A.

Measurement of MICs for this profile using the E-test showed that variations are only observable between *P. vulgaris*, *E. asburiae*, *E. gergoviae* and *P. mirabilis*. From 16 mg/L for *P. vulgaris* and *E. asburiae*, it increases to 64 mg/L for *E. gergoviae* and *P. mirabilis*. The fluctuations obtained with the MIC values for carbapenems in these microorganisms could be explained by the existence of two KPC variants between these identified enterobacterial species [31].

The second substrate and inhibition profile (P2) is associated with the carbapenemase identified in *E. gergoviae* isolated from wound pus. The enzymatic activity here shows several similarities with the P1 profile. The only difference is the decrease in enzyme activity in the presence of AMC. The MIC values do not differ from those obtained with *E. gergoviae* isolated from urine samples. This slight variation in MIC suggests that the same KPC is produced in the P1 profile by this microorganism in both urine and wound pus. The decrease in enzyme activity in the presence of clavunate may be due to insufficient ESBL production to contain all the suicide inhibitor molecules. The consequence is a decrease in the number of enzyme molecules available for substrate hydrolysis which would then lead to a decrease in enzyme activity.

The third substrate and inhibition profile (P3) is expressed by OXA-type carbapenemases (48 or 181) produced by *P. mirabilis* and *S. odorifera* 1 isolated from osteitis pus and blood samples respectively. This profile is characterised by enzymatic activity on CXC, decreased on the three carbapenems and not observed at all on AZT. EDTA has no discernible influence on this activity. All these characteristics are consistent with the description of a class D carbapenemase [35, 38]. Another observation on this profile is that the activity of the enzyme resumes on AZT in the presence of CTX. The resumption of enzyme activity on AZT in the presence of CTX illustrates the theory that the combination of two -lactams can be antagonistic if one of them is an -lactamase inducer. CTX would therefore induce the production of ESBLs that could hydrolyse AZT. This illustrates the fact that *P. mirabilis* and *S. odorifera* 1 are likely to produce inducible ESBLs in addition to OXAs. Analysis of the MICs obtained in these two species shows that there are no differences in the activity of this enzyme either at the level of the microorganisms or the samples. This suggests that the OXA produced by these microorganisms originates from the same plasmid that has migrated from one species to another [39].

The P4 profile is only found in *E. gergoviae* and *S. odorifera* 1 isolated from urine samples. It is characterised by an enzymatic activity on ETP and MRP but diminished with respect to IMP. The inhibitors clavunate and tazobactam have no effect on this enzymatic activity. This observation can be explained by the fact that these bacteria produce class B carbapenemases or, a combination of ESBL and chromosomal type A and/or B carbapenemases [16, 17]. The decrease in enzymatic activity in the presence of EDTA validates the hypothesis of the presence of a class B carbapenemase [38]. The combination of these observations leads us to believe that the genes coding for the synthesis of both class A and B carbapenemases, both chromosomal, are present in these bacteria. It is, therefore, the inhibition of class B carbapenemase by EDTA that would be at the origin of the decrease in enzymatic activity. In this context, the decrease in enzymatic activity would then be due to the reduction in the quantity of carbapenemases potentially active on carbapenems. In view of the above, it is possible that the bacteria *E. gergoviae* and *S. odorifera* 1 possess in their chromosomes both genes coding for the synthesis of class A and B carbapenemases. The MIC measurements for these microorganisms did not show any differences apart from that obtained with IMP (16 mg/L and 4 mg/L in *E. gergoviae* and *S. odorifera* 1 respectively). This difference in MICs can be explained by mutations that may occur in the amino acid sequence homology or by the level of production of one or the other of these carbapenemases.

The P5 substrate and inhibition profile is found in *P. mirabilis* isolated from urine samples and osteitis pus. This profile is characterised by enzymatic activity on all substrates except CAZ. This activity is maintained in the presence of all inhibitors. This suggests a most likely plasmid hyper production of KPC associated with cephalosporinase. The different profiles for this microorganism (P1 when derived from wound pus and P5 when derived from either urine samples or osteitis pus), although suspected of producing all the KPCs, would be the result of the difference in the enzyme that accompanies the production of these KPCs.

The P6 profile is found in *S. odorifera* 1 isolated from osteitis pus. It is characterised by an enzymatic activity towards ETP. This activity decreases on IMP and MRP. The presence of inhibitors has no visible effect on the enzymatic activity. The analysis of this P6 profile shows several similarities with the P3 profile. The same is true for the MIC values, which are close to those of the P3 profile. The great similarity observed between the P6 and P3 profiles suggest that *S. odorifera* 1 and *P. mirabilis*, both isolated from osteitis pus samples, produce carbapenemases of types OXA-48 or OXA-181. However, the increase in enzymatic activity observed with the P6 profile of *S. odorifera* 1 is thought to be the result of possible mutations in the OXA carbapenemases and the production of a cephalosporinase that activates the hydrolysis of CAZ [9].

The P7 profile expressed by *P. mirabilis* isolated from stools is characterised by an enzymatic activity on ETP and decreased on IMP and MRP substrates. The presence of clavunate shows inhibition of the enzymatic activity. This enzymatic activity, the extent of which varies from one carbapenemase to another, can be explained by the fact that it is the product of genes carried by the chromosomes [35, 40, 41]. The inhibition of the latter by clavunate validates the hypothesis of a class A carbapenemase.

The P8 profile identified in *S. odorifera* 1 isolated from stools shows enzymatic activity on ETP. This activity decreases on MRP and disappears on IMP. The inhibitor clavunate causes a loss of enzyme activity while EDTA has no effect on this activity. The MIC values show that the enzyme activity is distinct from one substrate to another. The fact that the enzyme activity is distinct on carbapenems and cephalosporin (CTX) sensitivity shows that this bacterium produces a chromosomal carbapenemase [35]. The inhibition of enzyme activity in the presence of clavunate supports the hypothesis of a class A carbapenemase [38]. Suspected carbapenemases may be SME, IMI-1 [40, 41]. The multiple similarities observed between the P7 and P8 profiles suggest that the carbapenemase produced in profile P8 may be a mutated form of that produced by *P. mirabilis* isolated from stool samples.

The P9 profile observed with the *C. braakii* microorganism isolated from stools is characterised by enzymatic activity on the ETP. This activity decreases with respect to MRP and disappears with respect to IMP and cephalosporins (CAZ, CTX). The presence of the inhibitor EDTA has no effect on the enzymatic activity contrary to clavunate and AZT which inhibit this activity. The strong similarity between the P9 and P8 profiles suggests that the same carbapenemase is mutated between *C. braakii* and *P. mirabilis* isolated from stool samples.

The P10 profile found in *E. gergoviae* isolated from stools always shows an enzymatic activity that varies from one carbapenem to another. With a few exceptions, this P10 profile is similar to the P9 profile. The observations show that the two profiles are similar and the few differences observed could reflect the presence of mutations in the genes producing these enzymes.

The last profile P11 is the fourth substrate and inhibition profile obtained with *E. gergoviae* isolated from blood. It is characterised by an enzymatic activity on ETP and IMP. This activity is diminished in the presence of MRP. The disappearance of this activity in the presence of EDTA indicates that the activity of this enzyme requires the presence of heavy metal [42]. No inhibition of the enzyme activity is observed with classical class A carbapenemase inhibitors. All these observations point to a class B carbapenemase [43]. The difficulty in typing this carbapenemase from this substrate and inhibition profile is the demarcation observed with other classical class B carbapenemases. This demarcation comes from the fact that the enzymatic activity here decreases towards MRP whereas class B carbapenemases exhibit enzymatic activity on all carbapenems [43]. The decreased enzymatic activity of this P11 profile on MRP can be explained by the presence of mutations in the primary amino acid sequence homology at the active site [44]. The presence of these mutations may be a consequence of being produced from integrons carrying 'cassette' genes from which several genes can be assembled [10]. The fact that this P11 profile shows activity on AZT and towards CAZ and CTX suggests the presence of an ESBL.

4. Conclusion

The objective of this work was to determine the types of carbapenemases moving around the city of Maroua in order to contribute to the development of a control strategy against the enterobacteria multidrug resistance. It was found that *Arizona, C. braakii, E. asburiae, E. gergoviae, P. mirabilis, P. vulgaris, S. ficaria* and *S. odorifera* 1 are the species of enterobacteria that produce carbapenemases in biological samples (urine, wound pus, osteitis pus, blood, stools). The carbapenemases identified are of the KPC, OXA and undetermined types. 11 different substrates and inhibition profiles are expressed by these microorganisms, some of which are able to produce two different classes of carbapenemases, others of producing a carbapenemase-BLSE or cephalosporinase combination. These 11 profiles have shown the difficulties of applying law of inhibition of these carbapenemases in situ in the context of probabilistic antibiotic therapy. This observation is valid whether the enterobacterium is identified or the biological medium of isolation is known.

Author details

Daoudou Bakari[1]*, Jojo Mbala Ekamba[1], Mamoudou Hamadou[1,2], Yaya Garga[1], Martin-Paul Baane[1,3] and Félicité Obono[2]

1 Faculty Science, Department of Biological Sciences, University of Maroua, Maroua, Cameroon

2 National Advanced School of Engineering of Maroua, University of Maroua, Maroua, Cameroon

3 HSWC-NSIF Biological Analysis Laboratory, Maroua, Cameroon

*Address all correspondence to: bkimou3@gmail.com

References

[1] Ebongue CO, Tsiazok DM, Mefo'o NPJ, Beyiha G, Adiogo D. Evolution de la résistance aux antibiotiques des entérobactéries isolées à l'Hôpital Général de Douala de 2005 à 2012. The Pan African Medical Journal. 2015;**20**: 227. DOI: 10.11604/pamj.2015.20.227.4770

[2] Gangoué-Pieboji J, Bedenic B, Koulla-Shiro S, Randegger C, Adiogo D, Ngassam P, et al. Extended spectrum beta lactamase producing enterobacteriaceae in Yaounde, Cameroon. Journal of Clinical Microbiology. 2005;**43**(7):3273-3277

[3] Lonchel CM, Meex C, Gangoué-Piéboji J, Boreux R, Assoumou MC, Melin P, et al. Proportion of extended-spectrum ß-lactamase producing enterobacteriaceae in community setting in Ngaoundere, Cameroon. BMC Infectious Diseases. 2012;**12**:53. DOI: 10.1186/1471-2334-12-53

[4] Fedama WM, Kingserbe L. Mécanismes de résistance aux antibiotiques des bactéries isolées des viandes de bœuf vendues dans les abattoirs et quelques points de ventes de la ville de Maroua (Mémoire DIPES II). Maroua, Cameroun: Université de Maroua; 2017. p. 48

[5] Naoupoul A. Mécanismes de résistance des bactéries isolées des yaourts vendus dans quelques sites de la ville de Maroua-Cameroun (Mémoire de Master). Ngaoundéré, Cameroun: Université de Ngaoundéré; 2017. p. 72

[6] Hamadou M, Daoudou B, Martin-Paul B, Mohamadou S, Roger DD. Inhibitory Effect of Methanolic and Methanolic-Aqueous Mixture Extract of Leaves of Plectranthus neochilus Schltr (Lamiaceae) and Bauhinia rufescens Lam (Fabaceae) on Two Strains of Enterobacteria Producing Beta-lactamases. Journal of Advances in Microbiology. 2020;**20**(7):11-20

[7] Hambate GV, Ndongo ES, Ombolo A, Loura B. Bacterial and physicochemical characterization of water from wells in urban areas of the District of Maroua III (Far North Cameroon). International Journal of Current Trends in Engineering& Technology. 2015;**1**(2): 43-50

[8] Cappuccino J, Sherman N. Microbiology, a Laboratory. 10th ed. Pearson Manual, Pearson Education; 2014. p. 567

[9] CASFM/EUCAST: Comité de l'Antibiogramme de la société Française de Microbiologie. Paris: Société Française de Microbiologie, Institut Pasteur; 2017. p. 127

[10] Dortet L, Poirel L, Nordmann P. Epidémiologie, détection et identification des entérobactéries productrices de carbapénémases. Feuillets de Biologie. 2013;**LIV**:312

[11] Nordmann P, Dortet L, Poirel L. Carbapenem resistance in *enterobacteriaceae*: Here is storm ! Trends in Molecular Medecine. 2012;**18**(5): 263-272

[12] Joly-Guillou ML. Intérêt du E-test dans le suivi de l'antibiothérapie. Réanimation. 2006;**15**(3):237-240. DOI: 10.1016/S1294-5501(06)70804-4

[13] Joffin JN, Leyral G. Microbiologie Technique. 2. Documentation Technique; Bordeaux [FRA]. CRDP Centre Régional de Documentation Pédagogique; 1998. p. 299

[14] Eveillard M, Le Fournis S, Prouteau C, Geslot A, Denos C, Kowalczyk JP, et al. Réservoir extra-digestif d'entérobactéries productrices de bêtalactamase à spectre élargi chez des patients non infectés: étude à partir d'une recherche systématique dans les urines et d'autres prélèvements à visée diagnostique. Annales de Biologie Clinique. 2012;**70**(6):717-723. DOI: 10.1684/abc.2012.0762

[15] Robin F, Gibold L, Bonnet R. Résistances naturelles et acquises aux β-lactamines chez les entérobactéries: Comment les identifier en pratique quotidienne? Revue Francophone des laboratoires. 2012;**2012**(445):47-58. DOI: 10.1016/S1773-035X(12)71676-3

[16] Nordmann P, Naas T, Poirel L. Global spread of Carbapenemase-producing Enteriobacteriaceae. Emerging Infectious Diseases. 2011;**17** (10):1791-1178

[17] Nordmann P, Poirel L, Walsh TR, Livermore DM. The emerging NDM carbapenemases. Trends in Microbiology. 2011;**19**(12):588-595

[18] Ouedraogo AS, Jean-Pierre H, Banus AL, Ouédraogo R, Godreuil S. Emergence et diffusion de la résistance aux antibiotiques en Afrique de l'Ouest: facteurs favorisants et évaluation de la ménace. Médécine et Santé Tropicale. 2017;**27**(2):147-154

[19] Wendt C, Schütt S, Dalpke AH, Konrad M, Mieth M, Trierwelier-Hauke B, et al. First outbreak of *Klebsiella pneumoniae* carbapenemase (KPC)-producing *K. pneumoniae* in Microorganismany. European Journal of Clinical Microbiology and Infectious Diseases. 2010;**29**:563-570

[20] Carattoli A. Plasmids and the spread of resistance. International Journal of Medical Microbiology. 2013;**303**:298-304

[21] Canchaya C, Fournous G, Chibani-Chennoufi S, Dillmann ML, Brüssow H. Phage as agent of lateral gene transfer. Current Opinion in Microbiology. 2003; **6**(4):417-424

[22] Guglielmini J, Quintais L, Garcillan-Barcia MP, De La Cruz F, Rocha EPC. The repertoire of ice in prokaryotes underscores the unity, diversity, and ubiquity of conjugaition. PLoS Genetics. 2011;**7**(8):e1002222. DOI: 10.1371/journal.pgen.1002222

[23] Attaiech L, Boughammoura A, Brochier-Armanet C, Allatif O, Peillard-Fiorente F, Edwards RA, et al. Silencing of natural transformation by RNA chaperone and a multitarget small RNA. Proceedings of the National Academy of Sciences USA. 2016;**113**(31):8813-8818

[24] El Bouamri MC, Arsalane L, Kamouni Y, Berraha M, Zouhair S. Evolution récente du profil épidémiologique des entérobactéries uropathogènes productrives de β-lactamases à spectre élargi à Marrakech Maroc. Progrès en Urologie. 2014;**24**:451-455

[25] Zeineb A. Profil épidémiologique des entérobactéries productrices de carbapènémases isolées au CHU IBN Sina-Rabat (Thése doctorat). Maroc: Université Mohammed V-Rabat; 2014. p. 174

[26] Bonnet R. β-lactamines et entérobactéries. In: Courvalin R, Leclercq R, Bingen E, editors. Antibiogramme. 2ème ed. ESKA, ASM Press; 2006. pp. 141-162

[27] Dortet L, Poirel L, Nordmann P. Rapid detection of carbapenemase-producing Pseudomonas spp. Journal of Clinical Microbiology. 2012;**50**(11):3773-3776

[28] Dortet L, Poirel L, Nordmann P. Rapid identification of carbapenemase

types in *Enterobacteriaceae* and Pseudomonas spp. by using a biochemical test. Antimicrobial Agents and Chemotherapy. 2012;**56**:6437-6440

[29] Walsh C. Antibiotics: Actions, Origins, Resistance. Washington, DC, USA: American Society for Microbiology Press; 2003. p. 344

[30] Nordmann P, Cuzon G, Naas T. The real threat of *Klebsiella pneumoniae* carbapenemase producing bacteria. Lancet Infectious Diseases. 2009;**9**(4):228-236

[31] Török ME, Moran E, Cooke FJ. Oxford Handbook of Infectious Diseases and Microbiology. 2nd ed. New York, USA: Oxford University Press; 2017. p. 909

[32] Poirel L, Potron A, Nordmann P. OXA-48-like carbapenemases: the phantom menace. Journal of Antimicrobial Chemotherapy. 2012;**67**(7):1597-1606

[33] Hamer CJ, Hall RM. IS26-mediated formation of transposons carrying antibiotic resistance genes. mSphere. 2016; **1**(2):1-8.e00038-1-8.e00016. DOI: 10.1128/mSphere.00038-16

[34] Carmeli Y, Akova M, Cornaglia G, Daikos GL, Garau J, Harbarth S, et al. Controlling the spread of carbapenemase-producing Gram-negatives: Therapeutic approach and infection control. Clinical Microbiology and Infection. 2010;**16**(2):102-111

[35] El-Herte RI, Kanj SS, Matar GM, Araj GF. The threat of carbapenem-resistant Enterobacteriaceae in Lebanon: An update on the regional and local epidemiology. Journal of Infection and Public Health. 2012;**5**(3):233-243

[36] Elhani D. Les bêta-lactamases à spectre étendu: le défi s'accentue. Annales de Biologie Clinique. 2012;**70** (2):117-140. DOI: 10.1684/abc.2012.0686

[37] Podolsky SH. The Antibiotic Era: Reform, Resistance, and the Pursuit of a Rational Therapeutics. North Charles Street, USA: Johns Hopkings University Press; 2015. p. 324

[38] Thomson KS. Extended-spectrum-beta-lactamase, AmpC, and Carbapenemase issues. Journal of Clinical Microbiology. 2010;**48**(4):1019-1025

[39] Evans BA, Amyes SGB. OXA β-lactamases. Clinical Microbioliogy Reviews. 2014;**27**(2):241-263

[40] Queenan AM, Bush K. Carbapenemases: The versatile beta-lactamases. Clinical Microbiology Reviews. 2007;**20**(3):440-458

[41] Queenan AM, Shang W, Schreckenberger P, Lolans K, Bush K, Quinn J. SME-3, anovel member of the *Serratia marcescens* SME family of carbapénème-hydrolysing β-lactamases. American Society for Microbiology. 2006;**10**(50):3485-3487

[42] Marchiaro P, Ballerini V, Spalding T, Cera G, Mussi MA, Moran-Barrio J, et al. A convenient microbiological assay employing cell-free extracts for the rapid characterization of Gram-negative carbapenemase producers. The Journal of Antimicrobial Chemotherapy. 2008; **62**:336-344

[43] Kon KMD, Rai M. Antibiotic Resistance: Mechanisms and New Antimicrobial Approaches. San Diego, CA, USA: Elsevier; 2016. p. 413

[44] Oelschlaeger P, Mayo SL, Pleiss J. Impact of remote mutations on metallo-β-lactamase substrate specificity: Implications for the evolution of antibiotic resistance. Protein Science. 2005;**14**:765-774

Chapter 7

Foodborne Pathogens of Enterobacteriaceae, Their Detection and Control

Ping Li, Han Jiang, Jiayi Xiong, Mengqi Fu, Xianpu Huang, Boxun Huang and Qing Gu

Abstract

Foodborne pathogens of *Enterobacteriaceae* including *Escherichia coli*, *Salmonella*, *Shigella, Yersinia*, etc., causes a great number of diseases and has a significant impact on human health. Here, we reviewed the prevalence, virulence, and antimicrobial susceptibility of *Enterobacteriaceae* belonging to 4 genera: *E. coli*, *Salmonella*, *Shigella,* and *Yersinia*. The routes of the pathogens' transmission in the food chain; the antimicrobial resistance, genetic diversity, and molecular epidemiology of the *Enterobacteriaceae* strains; novel technologies for detection of the bacterial communities (such as the molecular marker-based methods, Immunoaffinity based detection, etc.); and the controlling of the foodborne pathogens using chemical/natural compounds or physical methods (such as UV-C and pulsed-light treatment, etc.), is also summarized.

Keywords: foodborne pathogens, *Escherichia coli*, *Salmonella*, *Shigella*, *Yersinia*, detection and control

1. Introduction

Foodborne illness is the biggest health problem in the world. Due to unsanitary food processing methods, this situation is very serious in developing countries. Approximately 70% of diarrhea cases in developing countries are related to the consumption of contaminated food. An estimated 3.5 billion people have been infected, with 450 million people affected, most of them children [1]. There are many causes of foodborne illness, among which the most important are foodborne pathogens, including *E. coli (E. coli), Salmonella, Shigella,* and *Yersinia*. They can cause many diseases and have a significant impact on people's health and finance. *E. coli* is considered one of the main human foodborne pathogens. It is linked to a variety of acute and invasive human illnesses, and it is easy to spread across different ecosystems. *Salmonella* is a gram-negative, rod-shaped, flagellar facultative anaerobic bacteria belonging to the *Enterobacteriaceae* [2, 3]. *Salmonella* is divided into two categories: *Salmonella enterica* and *Salmonella bangri* [2, 3]. For *S. enterica*, more than 2600 sera have been isolated and described, many of which are pathogenic to humans and animals [2–4]. And *Shigella* is the third most common foodborne bacterial pathogen, according to the CDC. *Yersinia* also causes a range of foodborne illnesses with distinct characteristics in humans, ranging from asymptomatic carriers to hemorrhagic colitis and fatal typhoid fever.

In recent years, the detection of foodborne pathogens has developed rapidly. Many techniques such as PCR, nanotechnology, nucleic acid hybridization are widely used [5]. There are also many control methods for foodborne pathogens. In the present paper, we summarized the transmission, antimicrobial resistance, genetic diversity, and molecular epidemiology of the *Enterobacteriaceae* strains, and also novel technologies for detection and the controlling of the foodborne pathogens.

2. Transmission of pathogens in the food chain

Foodborne pathogens are transmitted through the food chain in many ways, such as insect transmission, fecal-oral transmission, food and water transmission, animals transmission, and so on. Some pathogens, such as *E. coli* or *Salmonella enteritidis* can be passed from animal hosts to people, but *Salmonella typhi* has no animal host and is highly harmful to humans.

Insects are considered to be carriers of foodborne pathogens. Their association with degradable substances and their endogenous and coexistence (with humans) are behavioral patterns that are particularly important for the ability of flies, cockroaches, and ants to transmit foodborne diseases. A study conducted in an ant colony in a Brazilian hospital found that several bacteria, including *E. coli* and *Salmonella,* were related to ants. Another study found cockroaches and several cockroach-related bacteria in several buildings in Spain, including *Salmonella* (hospitals), *E. coli* (hospitals, restaurants, companies, and grocery stores), and *Enterobacteria* (shops and food industry factories). In addition, an assessment of cockroaches gathered from hospitals, houses, grocery shops, and restaurants in the South Canary region of southwest India revealed that more than 4% of cockroaches tested positive for several *Salmonella* strains [6]. But existing understanding about the health dangers posed by flies and food is inadequate currently. Flies are at risk of transmitting foodborne pathogens because they have a bowel movement every 4 to 5 minutes during the day [7]. In general, houseflies can promote the spread of pathogens in four different ways: through body hair and surface, through the glandular hair on the feet, through the regurgitant rumen itus, and through the digestive tract [7]. Recently, some researchers have claimed that adult houseflies can spread their eggs and bacteria to food, so that these bacteria could be retransmitted to the first generation of adult flies [8]. Alexandre Lamas studied the bacterial populations of the Australian bush flies in three diverse places: cattle farms parking lots, metropolitan shopping malls, and a barbecue spot [9]. In the agricultural setting, the number of bacterial per fly was highest, whereas, it was lowest in the city [9]. Furthermore, multi-drug resistance was found in 94% of *Salmonella* isolates and 87% of *Shigella* isolates, suggesting that these flies might operate as food carriers for antimicrobial resistance transmission [10].

Water is well-known for its importance in the production, processing, and preparation of food. It is also a medium for the transmission of pathogens during food manufacturing [11]. The quantity of contamination in irrigation water determines pathogen survival, and the higher the degree of contamination, the better. They may survive outside of their human hosts for months to years before being transmitted to humans through water [12]. *E. coli* and *Salmonella* can leach through water or soil to the plant surface [13] and even *E. coli* O157:H7 can be absorbed by lettuce leaves. In addition, *E. coli* from livestock manure may persist for at least 5–6 months on soil or grassland, giving pathogens an excellent chance to infect other sources. In another research, *E. coli* O157:H7 could not only attach to the outer surface of radish seeds but also invade the inner tissues and stomata [14].

Many microorganisms that cause foodborne diseases can be transferred directly from animals to people. Mammals such as pigs and cattle are thought to host many foodborne pathogens, which are transmitted to humans either through direct contact with humans or by being processed into food for human consumption. *E. coli* is a typical element of the gut flora of humans and animals, and it is commonly found in poultry and wild animals. As a result, *E. coli* is one of the most likely infections to spread through food. The Shiga toxin-producing *E. coli* (STEC) strain is a serious foodborne pathogen that may be transmitted by consuming pig chow. From 334 pork samples collected from a South Korean slaughterhouse and retail market, 131 strains of *E. coli* were identified [15]. Simultaneously, *E. coli* was discovered in chickens. According to the Daily Mail, a food safety survey conducted in a supermarket in the UK found that 23 out of 99 chicken samples were infected with *E. coli*.

There are many key points where pathogens can infiltrate and jeopardize human food safety, such as the food itself, the surfaces of food preparation tools or food processors [16]. At each food processing or preparation facility location, a variety of factors may impact contamination and transmission. For example, microbial pathogens can be brought into the kitchen environment through commercial foods, cross-contamination of foods via kitchen equipment, or be reused due to inadequate cooking or storage [17, 18].

3. Antimicrobial resistance, genetic diversity and molecular epidemiology of the *Enterobacteriaceae* foodborne pathogens

3.1 *E. coli*

E. coli is one of the most common food-borne pathogens and may spread a variety of diseases through the food chain in different ecosystems. There are pathogenic and non-pathogenic strains of *E. coli*. Of these, pathogenic strains can cause a variety of intestinal diseases.

The original *E. coli* was sensitive to almost all antibacterial drugs [19], but multi-resistance of *E. coli* is now increasingly common. The resistance mechanism of *E. coli* includes the acquisition of encoding ultra-broad-spectrum β-lactamase (resistance to broad-spectrum cephalosporin), carbapenase (resistance to carbapenems), et al. The most common mechanism for the development of resistance in *E. coli* is the production of β-lactamase hydrolyzing β-lactamase antibiotics [20]. Ultra-broad-spectrum β-lactamases (ESBLs) are produced by mutations in β-lactamases and could be encoded by genes that effectively hydrolyze third and fourth-generation cephalosporins as well as monoclonal antibodies. However, β-lactamase inhibitors like clavulanate and tarmacadam can stop them [21]. Genes such as *aadA1*, *aadA2*, *mcr-1*, *crf*, and $bla_{\text{TEM-1}}$ are related to the drug resistance in *E. coli* (**Table 1**) [19].

The genetic diversity of *E. coli* is reflected not only at the individual level but also at the molecular level. Ramadan et al. [22] used Multilocus sequence typing (MLST) to explore the genetic diversity in *E. coli*, as indicated by the various distribution of *E. coli* lineages among different sources. It was found that a wide range of STs was found in chicken, human and beef isolates. And the most common STs isolated from chicken

Strain	Resistant phenotype	Resistance genes
Escherichia coli	Streptomycin/spectinomycin resistance	*aadA1, aadA2*
	Polymyxins resistance	*mcr-1*
	Fluorinated and nonfluorinated phenicols resistance	*crf*
	β-lactams resistance	bla_{TEM-1}
Salmonella	Beta-lactam resistance	*ampE*
	Macrolide resistance	*macB, macA*
	Aminoglycoside resistance	*aac6-I, acrD, acrD*
	Amidoalcohol (chloramphenicol) resistance	*mdfA, rarD*
	Amido alcohol (chloramphenicol) resistance	*gyrA, gyrB, parC, parE*、
	Other	*nfsA*
Shigella	Cephalosporins and Fluoroquinolones resistance	bla_{TEM-1}, bla_{CTX-M}, bla_{OXA-1}, bla_{SHV-12}
Yersinia	Tetracycline and minocycline resistance	*tetD, tetA*
	Ticarcillin and amoxicilin resistance	bla_{TEM-1B}
	Trimethoprim resistance	*dfrA14, drfA1*
	Sulfonamide resistance	*sul2*
	Chloramphenicol resistance	*catA2*

Table 1.
Resistance phenotype and resistance genes of the strain.

isolates differed significantly from human and beef isolates, which was consistent with previous research.

The genetic diversity of *E. coli* causes changes at the molecular level. Findlay et al. [23] revealed the cause of Urinary Tract Infection (UTI) was the direct sharing of *E. coli* between local farms and the local population. They found that the bla_{ctX-M} or $bla_{CMY\ 2}$ plasmid isolated from the farm *E. coli* isolates was almost identical to one of the three plasmids isolated from the urine of local people, and these three plasmids are found in almost all humans and animals on earth.

3.2 *Salmonella*

Salmonella is gram-negative bacteria. Based on the clinical presentation of the patient with their *Salmonella* infection, we usually identify them as typhoidal *Salmonella* and non-typhoidal *Salmonella* (NTS).

Salmonella has multidrug resistance because it is resistant to a variety of first-line antibiotics such as ampicillin, chloramphenicol and methicillin/sulfamethoxazole. Lu et al. [24] classified gene products by direct homology through functional annotation of the COG database. COG functional annotation was performed on 13 drug resistance genes of *Salmonella*, such as beta-lactam resistance and macrolide resistance. Also, they found that genes like *ampE*, *macB,* and *macA* are drug resistance genes in *Salmonella* (**Table 1**).

Salmonella is an important foodborne pathogen and its genetic diversity is of great significance for the prevention and control of the disease. Methods commonly used in genetic diversity research include serotyping and pulse electrophoresis typing, which are time-consuming and have poor traceability [25]. Zhang et al. [26] conducted multilocus sequence typing of 311 *salmonella* strains, and MLST typing results were divided into 26 ST types.

Molecular epidemiology has been used to document vector to human transmission and to investigate outbreaks of *Salmonellosis* in hospitals. *Salmonella* typing is epidemiologically important because it provides correlations between cases, foci, and between cases and food or other vectors, animals, regions, and periods. Riley et al. [27] studied an outbreak of enteritis in the northeastern United States in late 1981 caused by *Salmonella* Newport through commercially available raw beef. The outbreak strain is of the same serotype and is sensitive to most antibiotics. Plasmid analysis revealed two plasmids (3.7 and 3.4Md) of strains isolated from raw beef and patients with identical restriction profiles. Meanwhile, 45 percent of intestinal strains from New Jersey and Pennsylvania had the same plasmid profile. Through follow-up of patients, it was also found to be related to raw beef. Without molecular biological analysis, these cases would not be considered part of the outbreak.

3.3 *Shigella*

Shigella is the most common cause of diarrhoeal disease in humans worldwide, and its drug resistance is already a major public health burden. *Shigella* resistance tests have been reported in some areas of Shanxi Province, China. Of 474 strains, only 2 strains (0.5%) were sensitive to all 21 antimicrobial agents [28], 14 strains (3.0%) were co-resistant to the third-generation cephalosporins and fluoroquinolones. Wang et al. [29] found that $bla_{\text{TEM-1}}$, $bla_{\text{CTX-M}}$, $bla_{\text{OXA-1}}$, $bla_{\text{SHV-12}}$ are Cephalosporins and Fluoroquinolones resistance genes (**Table 1**).

Shigella is a common cause of diarrhea and death, particularly in children under the age of five. It is critical to investigate the genetic diversity of *Shigella*. Ei-Gendy et al. [28] isolated a total of 70 strains of *Shigella* from children younger than 5 years of age in Egypt, including 40 *Shigella dysenteriae* and 30 *Shigella boydii*. Among them, serotypes 7(30%), 2(28%), and 3(23%) accounted for the majority of *S. dysenteriae* isolates and 50% of *S. boydii* isolates were serotype 2.

Shigella is a common foodborne pathogen, and its molecular epidemiology is of great significance for the prevention and control of *Shigella*. Chen et al. [30] collected and typed 161 *Shigella* isolates obtained from Renai and adjacent townships from 1997 to 2000 using serological and PFGE techniques. The finding showed that the strain giving rise to foodborne illnesses remained the most common cause of *Shigellosis* during 4 years. Chen found that the percentage of these outbreak strain isolates among *Shigella flexneri* serotype 2a isolates recovered each year dropped. During this time, although several closely similar strains resembling outbreak strains have also emerged, they are far less transmissible and pathogenic than outbreak strains.

3.4 *Yersinia*

Yersinia pseudotuberculosis is the enteropathogen that causes gastrointestinal illnesses in people. Antibiotics that target gram-negative bacteria are typically effective against this species. However, the resistance to *Yersinia* is becoming more widespread. Three multi-drug-resistant (MDR) strains of *Y. pseudotuberculosis* were recovered from the environment in Russia and patients in France [31]. The resistance genes in *Yersinia* include *tetD*, *tetA* $bla_{\text{TEM-1B}}$, *dfrA14*, *drfA1*, *sul2* and *catA2*, etc., which are related to the tetracycline, minocycline, ticarcillin, amoxicillin and Trimethoprim resistance (**Table 1**).

The genetic diversity of *Yersinia pestis* is still mainly studied by typing. There have been many studies on the genetic diversity of *Yersinia*. Xu et al. [32] screened 102 *Y. pestis* isolates from Qinghai and 16 genotypes were identified by CRISPR (Clustered Regularly Interspaced Short Palindromic Repeat).

Yersinia is considered to be the pathogen of human intestinal diseases, and its molecular epidemiology is the focus of current research. The presence of a 70-kb virulence plasmid was required for the pathogenicity of *Y. pseudotuberculosis*, which was necessary for virulence. According to Fukushima [33], *Y. pseudotuberculosis* could produce a novel super antigenic toxin by chromosomal encoding, known as YPMa, YPMb or YPMc. It could also produce a pathogenicity island termed as HPI (high-pathogenicity island) or R-HPI (a right-hand part of the HPI with truncation in its left-hand part). All of these can contribute to its pathogenicity.

4. Novel technologies for detecting the pathogens

In recent years, the rapid detection of foodborne pathogens has developed rapidly. Molecular biology, nucleic acid hybridization, and other technologies have been highly valued and widely used in laboratory or factory production.

4.1 Nanoparticles in pathogen detection

Substances are manipulated at atomic, molecular, and supramolecular scales through nanotechnology ("nanotech"). Advances in manipulating these nanomaterials allow specific or non-specific binding of different biomolecules. The large specific surface area allows more biomolecules to be immobilized, thereby increasing the number of reaction sites that can be used to interact with the target species, which is one of the main advantages of biosensing using nanomaterials. In addition, nanomaterials have been widely used in 'label-free 'detection due to their excellent electronic and optical properties, and biosensors with enhanced sensitivity and improved response time have been developed [34].

Metal nanoparticles, especially gold and silver (5–110 nm in size) exhibit excellent properties, such as signal amplification, have potential application in various areas such as variable optical and electrical determinations. Gold nanoparticles (AuNPs) change the color aggregation from blue to red with the ability to scatter light, showing excellent chemical stability and electrical conductivity. AuNPs were used to detect Salmonella and *E. coli* O157: H7 organisms at 98.9 CFU/mL and 1–10 CFU/mL, respectively. Magnetic nanoparticles such as iron, nickel, and cobalt (size range of 1–100 nm) with electrical conductivity properties for utilization as a detection mean. Quantum dots (2–10 nm) were detected in *E. coli* O157:H7 10^3 CFU/mL through a semiconductor material consisting of semiconductor fluorescent nanonuclei (typically cadmium mixed with selenium or tellurium). Carbon nanotubes are formed by anisotropies of carbon-containing cylindrical graphene sheets. Multiwalled nanotubes (MWNTs, 2–100 nm) with photoluminescence and excellent electrical properties are composed of many concentrated single-walled nanotubes (SWNTs, 0.4–3 nm). A half conductance apparatus was used to monitor *E. coli* o157:h7 at 1 cell/mL restriction [35]. Thiol modified oligonucleotides covalently bound-based methods to gold nanoparticles are used as probes in various rapid detection ways. Due to its cost, functional chemistry is not so widespread. This method employs nonfunctional AuNPs to detect dsDNA and ssDNA [36].

4.2 Polymerase chain reaction (PCR)

Polymerase chain reaction (PCR) plays an important role in molecular methods in detecting foodborne pathogens. As early as 30 years ago, PCR, which was invented for the detection of single bacterial pathogens present in food by identifying specific target DNA sequences [37]. PCR works by amplifying specific target DNA sequences in a three-step cycle [38]. Firstly, single-stranded DNA was obtained from target double-stranded DNA by high-temperature denaturation. Then, deoxyribonucleic acid was lead on the backbone of DNA by adding specific primers and heat-resistant DNA polymerase in the polymerization process of DNA, so a new double-stranded DNA was synthesized. The amplified products of PCR were stained by ethidium bromide on electrophoretic gels [39]. PCR such as loop-mediated isothermal amplification (LAMP), multiplex PCR (mPCR) and RT-PCR, etc. is used to detect foodborne pathogens, including *E. coli* 157: H7, *S. aureus*, *Campylobacter jejuni*, *Salmonella* and *Shigella* [40]. Because of the advantage of high specificity, efficiency and easy operation, LAMP and mPCR are used quite frequently [41–47].

4.2.1 Loop-mediated isothermal amplification (LAMP)

Now, molecular diagnostic technologies based on nucleic acid amplification have been applied extensively in the detection regions, such as Loop-mediated isothermal amplification (LAMP) developed by Notomi [41–45]. Various confirmatory studies have been used to evaluate the feasibility of LAMP technology for microbial identification and diagnosis [42]. LAMP kits for detecting *Salmonella*, *E. coli*, and *Listeria monocytogenes* have been commercialized in the initial phase of development.

The loop-mediated isothermal amplification method offers several advantages: high sensitivity (2–5 orders of magnitude higher than conventional PCR methods); short reaction time (30–60 min can complete the reaction); no special instrumentation is required for clinical use; the operation is simple (whether DNA or RNA, the detection step is to mix the reaction liquid, enzyme, and template in a reaction tube, place in a water bath pot or incubator at 63°C for about 30 to 60 minutes, observe the results by the naked eye) [42–44]. There are also some disadvantages of the loop-mediated isothermal amplification method: high sensitivity, easy to form aerosol pollution once the lid is opened, combined with the current majority of domestic laboratories can not strictly partition, false-positive problems are relatively severe, so we strongly recommend using real-time turbidimeter during the development of the kit, do not open the reaction tube after the reaction. Primer design is more demanding, and some disease genes may not be amenable to the use of loop-mediated isothermal amplification methods [41–43].

4.2.2 Multiplex PCR (mPCR)

mPCR technology is more new-fashioned, which can simultaneously detect more pathogens than before, up to four or more pathogens [45–47]. Chen et al. simultaneously detected *S. enteritidis*, *S. flexneri*, and *E. coli* 157:H7 using five pairs of primers for invading protein (invA), 16S rDNA, invading plasmid antigen H (IPAH), Listeria hemolysin o (HlyA), and immunoglobulin (EAEA) genes [45]. The mPCR detection limit of mixed genomic DNA was 7.58×10^4 copies. Further improvements to mPCR by Gilmartin and O'Kennedy [46] promoted the process of a new GeXP PCR detection of four foodborne bacterial pathogens: *Salmonella*, *Yersinia*, *E. coli* 157:H7, and *Shigella*. The genome lab gene expression profiler (GeXP) gene analysis system can detect multiple pathogens in a single reaction with high throughput. Chimeric primers, universal primers and capillary electrophoresis with PCR products rather than agarose gel electrophoresis were involved in GeXP multiplex PCR amplification.

Synthesis of amplicons with universal tags by chimeric primers containing gene-specific sequences with universal tags at the 5′ end. Then, a universal primer will drive the remaining PCR reaction, which contains the same sequence of universal tags used by chimeric primers. Forward universal primer was covalently labeled with fluorescent dyes at the 5′ end for detection during capillary electrophoresis [47]. This method has higher sensitivity and is suitable for high-throughput analysis. Detection limits of Grignard PCR for *Salmonella*, *Yersinia*, *E. coli* 157:H7, and *Shigella*.

The characteristics of multiplex PCR are high efficiency, systematic and economic simplicity. High efficiency: a variety of pathogenic microorganisms in the same PCR reaction tube can be detected simultaneously, or multiple pathogens can be detected with multiple types of genes of interest. Systematic: mPCR is suitable for the detection of grouped pathogens. Economic simplicity: this will greatly economical of time, reagent and cost, and provide more accurate diagnostic information for clinical practice, because multiple pathogens are detected synchronously.

4.3 Nucleic acid hybridization technologies in pathogen detection

A general method of fluorescence in situ hybridization (FISH) using oligonucleotide probes of rRNA for nonmolecular technology. Probe lengths of 15 to 25 nucleotides labeled at the 5′ end were used for FISH. The specifically labeled cells were detected by an apparent fluorescence microscope. Rapid culture and independent detection of *Salmonella* were successfully performed using FISH combined with flow cytometry [48–50].

Line probe analysis (LIPA) is composed of oligonucleotide probes with specific oligonucleotides and nitrocellulose bands, which are connected by parallel lines along with the bands and discrete lines. The color change of hybridization results can be detected by vision. Innogenetics has produced several line probes for bacterial detection, such as *Escherichia coil*. The test results are consistent with those of antibiotics. Recently, 599 strains of *Escherichia coil* were improved and evaluated, and the sensitivity and specificity of the method were proved [51, 52].

Nielson et al. found a DNA analog called peptide nucleic acid (PNA) for detecting foodborne pathogens. This probe is more stable because PNA is not charged. In addition, PNA has a greater advantage in that it is relatively hydrophobic and easier to enter non-bacterial cells. PNA has higher specificity than DNA oligomer because the TM of the PNA probe is higher than that of its DNA probe. Theoretically, in addition to PNA and FISH, PNA can also replace DNA oligonucleotides to improve analytical performance [53, 54].

5. Controlling of the *Enterobacteriaceae* foodborne pathogens

At present, food pollution and poisoning caused by foodborne pathogens have attracted extensive attention. In the food industry, technologies such as irradiation, pulsed light treatment, microwave sterilization, slightly acid electrolytic water and fumaric acid treatment, algae extract treatment, *Bacillus* antimicrobial peptide treatment is usually used to control foodborne pathogens.

5.1 Irradiation

In more and more countries, ionizing radiation processing is the most common method of food purification, and in the short run, a growing number of radiation-purified foods are presumed to be approved for production. It is a secure, smart, environmentally clean, and energy-efficient process, and it is especially valuable as a purification process for the final product. Due to the availability of irradiation in handling packaged foods, irradiation is regarded by most food safety officers and scientists as an effective critical control point in the processing of meat and poultry hazard analysis and critical control point (HACCP) system.

The high-energy photons or free radicals generated by ionizing radiation can break the DNA chain and generate reactive oxygen free radicals, and can also cause protein denaturation and cell membrane damage. Hesham reported that an irradiation dose of 4 kGy can effectively control the bacterial pathogens in meat by destroying *Salmonella*, significantly reducing *E. coli* [55]. They found the number of *Enterococcus faecalis* and *Enterobacteriaceae* was reduced by more than 1.8 log units and 5 log units, respectively, when treated with 4 kGy of irradiation, and no *Salmonella* was detected in the meat samples [55], which could prolong the cold storage shelf life without any significant impact on the sensory quality of meat.

5.2 Pulsed-light treatment

Nucleic acids are easily destroyed by pulsed light (PL). Pyrimidine bases form dimers the DNA of bacteria, viruses, and other pathogens through photochemical intervention and block DNA replication, and if there is not enough repair mechanism, it will ultimately lead to the death of microorganisms [56]. Xu et al. [57] investigated the inactivation effect of PL on *Salmonella* and *E. coli* in fresh raspberries. It was found that the pulsed light treatment of 28.2 J/cm^2 for 30 s could reduce them by 4.5 and 3.9 lgCFU/g, respectively. However, considering the adverse effects on raspberry color and ground, the recommended dosage of PL is 5.0 J/cm^2. Rajkovic et al. [58] found that PL can kill *E. coli* in meat products, but the sterilization effect becomes worse with the extension of pulse interval. Ozer et al. [59] used pulsed ultraviolet light to treat *E. coli* on the surface of seafood. The results showed that the irradiation distance was 5 cm and the treatment time was 30 s, reducing 0.86 lgCFU/g; When the irradiation distance was 8 cm and treated for 60 s, 1.09 lgCFU/g was reduced [60]. This shows that under the condition of a long irradiation distance, the sterilization rate can be improved by prolonging the treatment time, but the surface temperature of the sample increases significantly with the extension of the treatment time.

However, in the sterilization process of fruits and vegetables, if the PL intensity is too high, due to the effect of PL on protein structure, it will improve the activity of polyphenol oxidase (PPO) to a certain extent and cause browning [61]. In the process of meat sterilization, PL has a poor sterilization effect on uneven surfaces [62], and the sterilization only stays on the surface.

5.3 Microwave sterilization

Microwave sterilization is that microwave constantly changes the direction of electromagnetic field, changes the ion and electron density around microbial cell membrane, destroy the permeability of cell membrane, lead to protein degeneration in cells, destroy cell metabolism, and microbial death [63].

De La Vega-Miranda observed that under 950 W water-assisted microwave treatment, *Salmonella typhimurium* on pepper and coriander foliage decreased by 5.12 log and 4.45 log after being treated at 63°C for 25 s and 10 s, respectively, and finally reached 3×10^8 CFU/g [64]. The sterilization effect of microwave sterilization under the same conditions (power and temperature) varies due to different objects. The high-voltage pulsed electric field sterilization technology to treat liquid food shows that it can effectively eliminate *E. coli*, *Salmonella*, *E. coli* O157:H7, et al. reaching the level of pasteurization. The cold source plasma has a significant sterilization effect on *Salmonella* and *B. subtilis* in pepper, and the cavitation jet technology also has a significant sterilization effect on *E. coli* and *K. pneumonia*.

5.4 Slightly acidic electrolyzed water and fumaric acid

Slightly acidic electrolyzed water (SAcEW) is a type of EW and promising sanitizer for food products. Effects of SAcEW combination with other chemical disinfectants on the ideal bactericidal efficacy of foods. Organic acids can inactivate foodborne pathogens, and show stronger bactericidal effects in organic acids used in meat antibacterial agents.

Ahmad found that a single treatment and combined treatment of fresh meat with micro-electrolyzed water or fumaric acid can reduce *E. coli* and *S. Typhimurium* in meat [65]. The efficacy of *Salmonella* and study the quality guarantee period and organoleptic quality of the meat during conserve at 5°C and 12°C. The inoculated meat samples were soaked for 5 min in each treatment, with or without gentle heating. Compared with other treatments, SAcEW +0.6% FA 40°C 5 min had a stronger bactericidal effect on fresh meat and significantly lessened *E. coli* and *Salmonella* respectively reduced 2.34 and 2.88 logCFU/g. This combined treatment made the natural bacteria (TBC) lag time of meat stored at 5°C longer. Compared with the untreated meat, the treatment of combined extended the quality guarantee period of meat by 8 days and 6–7 days when respectively stored at 5°C and 12°C. The study has shown that the combined treatment of SAcEW +0.6% FA has the potential as a new way to improve the microbial security and quality of fresh meat [65].

5.5 Other technologies for controlling the *Enterobacteriaceae* foodborne pathogens

Recent studies have shown that some biological macromolecules can also be used to control foodborne pathogens of *Enterobacteriaceae*, such as *Bacillus* antimicrobial peptides and algae extracts. Chen et al. [66] found that *Bacillus* antimicrobial peptides can be applied to the control of food-borne pathogens in seafood, but there are still many key issues that need to be further studied, especially the effect of *Bacillus* antimicrobial peptides and their main active ingredients on common foodborne pathogens in seafood antibacterial effect; the relationship between the dose of *Bacillus* antimicrobial peptides and the survival and production of toxins in complex food environments; key issues such as the mode of action of bacillus antimicrobial peptides at the cellular and molecular levels on pathogenic bacteria.

Algae is a multifaceted natural substrate that contains a wide range of bioactive compounds. Antibacterial, analgesic, and antioxidant properties of phytosterols isolated from different algae have been demonstrated. Brown algae fucoidans and green algea ulvans both have antibacterial capacities. The most potent chemicals against *E. coli* are carvacrol and thymol [67]. Algae and alga extracts have also been reported as having the ability to enhance food quality when used as feedstock, as well as assisting in the management of microbial contamination in fish farms [68]. Nowadays, algae-rich foods have emerged, food safety, functional food, and non-traditional diet are worthy of attention [69–71]. Algae are a kind of available resource for new bioactive molecules. Therefore, Algae have great potential for application in

Foodborne pathogens	Treatments	Results/Activity	Reference
Escherichia coli	4 kGy dose of radiation	Reduce >5 log units	[66]
	Slightly acidic electrolyzed water and fumaric acid	Reduce 2.34 log CFU/g	[65]
	Brown Algae Methanol Extract	Sensitive	[67]
	Phage cocktail	Spraying the phage mixture resulted in a 4.5 log CFU reduction after 2 h	[72]
	Phage DT1 and DT6	100% reduction in CFU/ml within an hour	[73]
	Lactobacillus acidophilus A4	Anti-adhesive/ Antibiofilm	[74]
	L. acidophilus La-5	Anti-quorum sensing	[75]
	Carvacrol, thymol, trans-cinnamaldehyde	Antibiofilm Reduced expression of virulence genes	[76]
	Surface-layer protein extract	Anti-adhesive	[77]
	Resveratrol	Antibiofilm	[78]
	Microwave radiation	Elimination of the superficial	[79]
Salmonella	4 kGy dose of radiation	Not detected	[55]
	Water-assisted microwave heating	5.12 log reduction	[64]
	slightly acidic electrolyzed water and fumaric acid	Reduce 2.88 log CFU/g	[65]
	Brown Algae Methanol Extract	Sensitive	[67]
	Phage cocktail	Using MOI 5 leads to about 4.4 log reductions	[60]
	Phage F01-E2	The CFU of turkey cooked meat and chocolate milk was reduced by 5 log, and the CFU of hot dog was reduced by 3 log	[80]
	Phage cocktail PC1	More than 99% reduction in CFU at MOI 10 or above	[81]
	Bifidobacterium lactis Bb12/*Lactobacillus rhamnosus* LGG	Anti-adhesive	[82]
	E. coli Nissle	Anti-invasive	[83]
	T315 compound	Antibiofilm	[84]
	Methylthioadenosine	Reduced motility Anti-invasive	[85]
	Microwave radiation	Theoretical complete inactivation	[86]
Shigella	Phage cocktail	About 4 log reduction	[87]

Foodborne pathogens	Treatments	Results/Activity	Reference
	Containing six novel *Shigella* specific phages	About 99% decrease	[88]
Yersinia	*Yersinia enterocolitica* phages	Decreasing by 1–3 logs on food samples	[89]
	Bacteriophage specific to serotype O1 *Yersinia ruckeri* (φNC10)	Polysaccharide Depolymerase activity capable of degrading *Y. ruckeri* O1-LPS	[90]

Table 2.
Controlling of the Enterobacteriaceae *foodborne pathogens.*

controlling foodborne pathogens [70]. Algae may be used as fresh food preservatives, active packaging, or antifouling and biofilm inhibitors based on the above advantages. To maximize the advantages of algae and algae compounds in food safety, attractive sensory characteristics should be pursued shortly (**Table 2**).

6. Conclusion

A plenty number of studies have been confirmed that foodborne pathogens of *Enterobacteriaceae* and their resistance genes can not only remain in animal husbandry and related environment but also transmitted to human beings through the food chain or other ways, causing a major threat to public health. Also, it has been highlighted how much important are novel technologies for the detection of foodborne pathogens (such as molecular marker-based methods, immunoaffinity-based detection, etc.). In addition, chemical/natural compounds or physical methods (such as UV-C and pulsed-light treatment, etc.) play key roles in the prevention of foodborne pathogen growth and diffusion. As one of the causes of foodborne diseases of global concern, foodborne pathogens should be controlled by countries and organizations around the world through the establishment of policies and food safety management systems.

Acknowledgements

This project was funded by the Natural Science Foundation of Zhejiang Province (LR22C200005), the Key Research and Development Program of Zhejiang Province (2020C04002) and the Science and Technology Project of Zhejiang Province (LGN22C200013).

DOI: http://dx.doi.org/10.5772/intechopen.102086

Author details

Ping Li[1]*, Han Jiang[2], Jiayi Xiong[1], Mengqi Fu[1], Xianpu Huang[1], Boxun Huang[1] and Qing Gu[1]

1 Key Laboratory for Food Microbial Technology of Zhejiang Province, College of Food Science and Biotechnology, Zhejiang Gongshang University, Hangzhou, China

2 Key Laboratory of Marine Food Quality and Hazard Controlling Technology of Zhejiang Province, College of Life Sciences, China Jiliang University, Hangzhou, Zhejiang, China

*Address all correspondence to: ping-biology@outlook.com

References

[1] Bintsis T. Foodborne Pathogens. Kastoria, Greece: Department of International Trade, TEI of West Macedonia; 2017

[2] Laxminarayan R, Duse A, Wattal C, et al. Antibiotic resistance-the need for global solutions. The Lancet Infectious Diseases. 2013;**13**(**12**):1057-1098

[3] Jajere SM. A review of *Salmonella* enterica with particular focus on the pathogenicity and virulence factors, host specificity and adaptation and antimicrobial resistance including multidrug resistance. Veterinary World. 2019;**12**:504-521

[4] Eng S-K, Pusparajah P, Ab Mutalib N-S, Ser H-L, Chan K-G, Lee L-H. *Salmonella*: A review on pathogenesis, epidemiology and antibiotic resistance. Frontiers in Life Science. 2015;**8**:284-293

[5] Bertelloni F, Tosi G, Massi P, Fiorentini L, Parigi M, Cerri D, et al. Some pathogenic characters of paratyphoid *Salmonella* enterica strains isolated from poultry. Asian Pacific Journal of Tropical Medicine. 2017;**10**:1161-1166

[6] Mezal EH, Sabol A, Khan MA, Ali N, Stefanova R, Khan AA. Isolation and molecular characterization of *Salmonella* enterica serovar Enteritidis from poultry house and clinical samples during 2010. Food Microbiology. 2014;**38**:67-74

[7] Song XY, Ming SX, Wang HW, He JX. Research progress in rapid biological detection of food-borne pathogens. Journal of Food Safety and Quality. 2021;**12**(**21**):8582-8589. DOI: 10.19812/j.cnki.jfsq11-5956/ts.2021.21.040

[8] Stein RA, Chirilã M. Routes of Transmission in the Food Chain. New York, NY: New York University School of Medicine, United States and Department of Natural Sciences; 2020

[9] Lamas A, Regal P, Vázquez B, Miranda JM, Franco CM, Cepeda A. Transcriptomics: A powerful tool to evaluate the behavior of foodborne pathogens in the food production chain. Food Research International. 2019;**125**:108543. DOI: 10.1016/j.foodres.2019.108543

[10] Peacock BD. Tracing pathogens in the food chain. International Journal of Dairy Technology. 2011;**64**(4):256-288

[11] Kalyoussef S, Feja KN. Foodborne illnesses. Advances in Pediatrics. 2014;**61**(**1**):287-312. DOI: 10.1016/j.yapd.2014.04.003

[12] Steele M, Odumeru J. Irrigation water as source of foodborne pathogens on fruit and vegetables. Journal of Food Protection. 2004;**67**:2839-2849

[13] Franz E, Visser AA, Van Diepeningen AD, Klerks MM, Termorshuizen AJ, Van Bruggen AH. Quantification of contamination of lettuce by GFP-expressing *Escherichia coli* O157:H7 and *Salmonella enterica* serovar Typhimurium. Food Microbiology. 2007;**24**:106-112

[14] Avery SM, Moore A, Hutchison ML. Fate of *Escherichia coli* originating from livestock faeces deposited directly onto pasture. Letters in Applied Microbiology 2004; **38**, 355-359.

[15] Heo EJ, Ko EK, Kang HJ, Kim YJ, Park HJ, Wee SH, et al. Prevalence and antimicrobial characteristics of Shiga toxin-producing *Escherichia coli* isolates from pork in Korea. Foodborne

Pathogens and Disease. 2020;**17**(**10**):602-607. DOI: 10.1089/fpd.2019.2760

[16] Todd EC, Greig JD, Bartleson CA, Michaels BS. Outbreaks where food workers have been implicated in the spread of foodborne disease. Part 6. Transmission and survival of pathogens in the food processing and preparation environment. Journal of Food Protection. 2009;**72**:202-219

[17] Hall G, Vally H, Kirk M. Foodborne illnesses: Overview. International Encyclopedia of Public Health. 2008;**17**:638-653

[18] Vogt RL, Dippold L. *Escherichia coli* O157:H7 outbreak associated with consumption of ground beef, June–July 2002. Public Health Reports. 2005;**120**:174-178

[19] Poirel L, Madec JY, Lupo A, Schink AK, Kieffer N, Nordmann P, et al. Antimicrobial Resistance in *Escherichia coli*. Microbiology Spectrum. 2018;**6**(**4**):1-27. DOI: 10.1128/microbiolspec.ARBA-0026-2017

[20] Pitout JD, Thomson KS, Hanson ND, Ehrhardt AF, Moland ES, Sanders CC. beta-lactamases responsible for resistance to expanded-spectrum cephalosporins in *Klebsiella pneumoniae*, *Escherichia coli*, and *Proteus mirabilis* isolates recovered in South Africa. Antimicrobial Agents and Chemotherapy. 1998;**42**(**6**):1350-1354. DOI: 10.1128/AAC.42.6.1350

[21] Fernandes R, Amador P, Oliveira C, Prudêncio C. Molecular characterization of ESBL-producing *Enterobacteriaceae* in northern Portugal. Scientific World Journal. 2014;**2014**:782897. DOI: 10.1155/2014/782897

[22] Ramadan H, Jackson CR, Frye JG, Hiott LM, Samir M, Awad A, et al. Antimicrobial resistance, genetic diversity and multilocus sequence typing of *Escherichia coli* from humans, retail chicken and ground beef in Egypt. Pathogen. 2020;**9**(**5**):357. DOI: 10.3390/pathogens9050357

[23] Findlay J, Mounsey O, Lee WWY, Newbold N, Morley K, Schubert H, et al. Molecular epidemiology of *Escherichia coli* producing CTX-M and pAmpC β-lactamases from dairy farms identifies a dominant plasmid encoding CTX-M-32 but No evidence for transmission to humans in the same geographical region. Applied and Environmental Microbiology. 2020;**87**(**1**):e01842. DOI: 10.1128/AEM.01842-20

[24] Lu JY, Z. K., Zhao WW, Li R. Genome-wide screening of resistant genes and resistant plasmids in Salmonella. Journal of Wuhan Polytechnic University. 2020;**39**: 20-25.

[25] Zhao C, Guo SY, Li LL, Liu YJ, Wang YM, Chang WS, et al. Study on MLST and serological typing and distribution of Salmonella from animals in Shandong Province. Chinese Journal of Zoonoses. 2017;**33**:793-799

[26] Zhang L. Serotype, Drug Resistance and Molecular Epidemiology of *Salmonella* from Chickens. The Chinese Veterinary Drug Supervision; 2021. DOI: 10.27645/dcnki. Gzsys.2021.000004

[27] Riley LW et al. The Journal of Infectious Diseases. 1983;**148**:12-17

[28] El-Gendy AM, Mansour A, Weiner MA, Pimentel G, Armstrong AW, Young SY, et al. Genetic diversity and antibiotic resistance in *Shigella* dysenteriae and *Shigella* boydii strains isolated from children aged <5 years in Egypt. Epidemiology and Infection. 2012;**140**(**2**):299-310. DOI: 10.1017/S0950268811000525

[29] Wang Y, Ma Q, Hao R, Zhang Q, Yao S, Han J, et al. Antimicrobial resistance and genetic characterization of Shigella spp. in Shanxi Province, China, during 2006-2016. BMC Microbiology. 2019;**19**(**1**):116. DOI: 10.1186/s12866-019-1495-6

[30] Chen JH, Chiou CS, Chen PC, Liao TL, Liao TL, Li JM, et al. Molecular epidemiology of *Shigella* in a Taiwan township during 1996 to 2000. Journal of Clinical Microbiology. 2003;**41**(7):3078-3088. DOI: 10.1128/JCM.41.7.3078-3088.2003

[31] Cabanel N, Galimand M, Bouchier C, Chesnokova M, Klimov V, Carniel E. Molecular bases for multidrug resistance in *Yersinia* pseudotuberculosis. International Journal of Medical Microbiology. 2017;**307**(7):371-381. DOI: 10.1016/j.ijmm.2017.08.005

[32] Xu X, Cui Y, Xin Y, Yang X, Zhang Q, Jin Y, et al. Genetic diversity and spatial-temporal distribution of *Yersinia* pestis in Qinghai plateau, China. PLoS Neglected Tropical Diseases. 2018;**12**(**6**):e0006579. DOI: 10.1371/journal.pntd.0006579

[33] Fukushima H. Molecular epidemiology of *Yersinia pseudotuberculosis*. Advances in Experimental Medicine and Biology. 2003;**529**:357-358. DOI: 10.1007/0-306-48416-1_70

[34] Gilmartin N, O'Kennedy R. Nanobiotechnologies for the detection and reduction of pathogens. Enzyme and Microbial Technology. 2012;**50**(**2**):87-95. DOI: 10.1016/j.enzmictec.2011.11.005

[35] Rantala A, Rizzi E, Castiglioni B, de Bellis G, Sivonen K. Identification of hepatotoxin-producing cyanobacteria by DNA-chip. Environmental Microbiology. 2008;**10**(**3**):653-664. DOI: 10.1111/j.1462-2920.2007.01488.x

[36] Rho S, Kim SJ, Lee SC, Chang JH, Kang HG, Choi J. Colorimetric detection of ssDNA in a solution. Current Applied Physics. 2009;**9**(**2**):534-537

[37] Chen J, Griffiths MW. Salmonella detection in eggs using lux+ bacteriophages. Journal of Food Protection. 1996;**59**(**9**):908-914

[38] Chen SH, Wu VC, Chuang YC, Lin CS. Using oligonucleotide-functionalized Au nanoparticles to rapidly detect foodborne pathogens on a piezoelectric biosensor. Journal of Microbiological Methods. 2008;**73**(**1**):7-17. DOI: 10.1016/j.mimet.2008.01.004

[39] Rantala A, Rizzi E, Castiglioni B, de Bellis G, Sivonen K. Identification of hepatotoxin-producing cyanobacteria by DNA-chip. Environmental Microbiology. 2008;**10**(**3**):653-664. DOI: 10.1111/j.1462-2920.2007.01488.x

[40] Zhao J, Jedlicka SS, Lannu JD, Bhunia AK, Rickus JL. Liposome-doped nanocomposites as artificial-cell-based biosensors: Detection of listeriolysin O. Biotechnology Progress. 2006;**22**(**1**):32-37. DOI: 10.1021/bp050154o

[41] Velusamy V, Arshak K, Korostynska O, Oliwa K, Adley C. An overview of foodborne pathogen detection: in the perspective of biosensors. Biotechnology Advances. 2010;**28**(**2**):232-254. DOI: 10.1016/j.biotechadv.2009.12.004

[42] Mandal PK, Biswas AK, Choi K, Pal UK. Methods for rapid detection of foodborne pathogens: An overview. American Journal of Food Technology. 2011;**6**:87-102. DOI: 10.3923/ajft.2011.87.102

[43] Zhao X, Lin CW, Wang J, Oh DH. Advances in rapid detection methods

for foodborne pathogens. Journal of Microbiology and Biotechnology. 2014;**24**(**3**):297-312. DOI: 10.4014/jmb.1310.10013

[44] Zhou B, Xiao J, Liu S, Yang J, Wang Y, Nie F, et al. Simultaneous detection of six food-borne pathogens by multiplex PCR with GEXP analyzer. Food Control. 2013;**32**:198-204. DOI: 10.1016/j.foodcont.2012.11.044

[45] Chen J, Tang J, Liu J, Cai Z, Bai X. Development and evaluation of a multiplex PCR for simultaneous detection of five foodborne pathogens. Journal of Applied Microbiology. 2012;**112**(**4**):823-830. DOI: 10.1111/j.1365-2672.2012.05240.x

[46] Gilmartin N, O'Kennedy R. Nanobiotechnologies for the detection and reduction of pathogens. Enzyme and Microbial Technology. 2012;**50**(**2**):87-95. DOI: 10.1016/j.enzmictec.2011.11.005

[47] Yang L, Li Y. Simultaneous detection of *Escherichia coli* O157:H7 and *Salmonella Typhimurium* using quantum dots as fluorescence labels. The Analyst. 2006;**131**(**3**):394-401. DOI: 10.1039/b510888h

[48] Bisha B, Brehm-Stecher BF. Combination of adhesive-tape-based sampling and fluorescence in situ hybridization for rapid detection of Salmonella on fresh produce. Journal of Visualized Experiments. 2010;**44**:2308. DOI: 10.3791/2308

[49] Poppert S, Essig A, Marre R, Wagner M, Horn M. Detection and differentiation of chlamydiae by fluorescence in situ hybridization. Applied and Environmental Microbiology. 2002;**68**(8):4081-4089. DOI: 10.1128/AEM.68.8.4081- 4089.2002

[50] Trebesius K, Adler K, Vieth M, Stolte M, Haas R. Specific detection and prevalence of helicobacter heilmannii-like organisms in the human gastric mucosa by fluorescent in situ hybridization and partial 16S ribosomal DNA sequencing. Journal of Clinical Microbiology. 2001;**39**(4):1510-1516. DOI: 10.1128/JCM.39.4.1510-1516.2001

[51] Rho S, Kim SJ, Lee SC, Chang JH, Kang HG, Choi J. Colorimetric detection of ssDNA in a solution. Current Applied Physics. 2009;**9**(2):534-537

[52] Mijs W, De Vreese K, Devos A, Pottel H, Valgaeren A, Evans C, et al. Evaluation of a commercial line probe assay for identification of mycobacterium species from liquid and solid culture. European Journal of Clinical Microbiology & Infectious Diseases. 2002;**21**(11):794-802. DOI: 10.1007/s10096-002-0825-y

[53] Yang L, Bashir R. Electrical/electrochemical impedance for rapid detection of foodborne pathogenic bacteria. Biotechnology Advances. 2008;**26**(2):135-150. DOI: 10.1016/j.biotechadv.2007.10.003

[54] Lin YH, Chen SH, Chuang YC, Lu YC, Shen TY, Chang CA, et al. Disposable amperometric immunosensing strips fabricated by Au nanoparticles-modified screen-printed carbon electrodes for the detection of foodborne pathogen *Escherichia coli* O157:H7. Biosensors & Bioelectronics. 2008;**23**(**12**):1832-1837. DOI: 10.1016/j.bios.2008.02.030

[55] Hesham M. Badr. Use of irradiation to control foodborne pathogens and extendthe refrigerated market life of rabbit meat. Meat Science. 2004;**67**:541-548. DOI: 10.1016/j.meatsci.2003.11.018

[56] MacGregor SJ, Rowan NJ, Mcllvaney L, Anderson JG, Fouracre RA, Farish O. Light inactivation of food-related pathogenic bacteria using a

pulsed power source. Letters in Applied Microbiology. 1998;**27**(**2**):67-70. DOI: 10.1046/j.1472-765x.1998.00399.x

[57] Xu W, Wu C. The impact of pulsed light on decontamination, quality, and bacterial attachment of fresh raspberries. Food Microbiology. 2016;**57**:135-143. DOI: 10.1016/j.fm.2016.02.009

[58] Rajkovic A, Tomasevic I, Smigic N, et al. Pulsed UV light as an intervention strategy against *Listeria monocytogenes* and *Escherichia coli* O157: H7 on the surface of a meat slicing knife. Journal of Food Engineering. 2010;**100**(**3**):446-451. DOI: 10.1016/j.jfoodeng.2010.04.029

[59] Ozer NP, Demirci A. Efficacy of pulsed-UV light treatment inactivating *Escherichia coli* O157: H7 and *Listeria monocytogenes* on raw salmon. In: 2004 ASAE Annual Meeting. Vol. 1. Ottawa: American Society of Agricultural and Biological Engineers; 2004

[60] El-Dougdoug NK, Cucic S, Abdelhamid AG, Brovko L, Kropinski AM, Griffiths MW, et al. Control of *Salmonella* Newport on cherry tomato using a cocktail of lytic bacteriophages. International Journal of Food Microbiology. 2019;**293**:60-71. DOI: 10.1016/j.ijfoodmicro.2019.01.003

[61] Ingrid A, Gemma O, Olga M, Robert S. Impact of pulsed light treatments on quality characteristics and oxidative stability of fresh-cut avocado. LWT--Food Science and Technology. 2014;**59**(**1**):320-326. DOI: 10.1016/j.lwt.2014.04.049

[62] Choi MS, Cheigh CI, Jeong EA, Shin JK, Chung MS. Nonthermal sterilization of *Listeria monocytogenes* in infant foods by intense pulsed-light treatment. Journal of Food Engineering. 2010;**97**(**4**):504-509. DOI: 10.1016/j.jfoodeng.2009.11.008

[63] Shahla A, Masoud S, Azam M, Fatemeh M. PbTe nanostructures: Microwave-assisted synthesis by using lead Schiff-base precursor, characterization and formation mechanism. Comptes Rendus Chimie. 2013;**16**(**9**):778-788. DOI: 10.1016/j.crci.2013.03.017

[64] De La Vega-Miranda B, Santiesteban-López NA, López-Malo A, Sosa-Morales ME. Inactivation of *Salmonella Typhimurium* in fresh vegetables using water-assisted microwave heating. Food Control. 2012;**26**(**1**):19-22. DOI: 10.1016/j.foodcont.2012.01.002

[65] Ahmad R, Charles N, Gwang-Hee K, Deog-Hwan O. Combined effects of slightly acidic electrolyzed water and fumaric acid on the reduction of foodborne pathogens and shelf life extension of fresh pork. Food Control. 2015;**47**:277-284. DOI: 10.1016/j.foodcont.2014.07.019

[66] Chen Y, Zhang YF, Wang XH, Ling JQ, He GH, Shen LR. Antibacterial activity and its mechanisms of a recombinant Funme peptide against Cronobacter sakazakii in powdered infant formula. Food Research International. 2019;**116**:258-265. DOI: 10.1016/j.foodres.2018.08.030

[67] Silva A, Silva SA, Lourenço-Lopes C, Jimenez-Lopez C, Carpena M, Gullón P, et al. Antibacterial use of macroalgae compounds against foodborne pathogens. Antibiotics. 2020;**9**(**10**):712. DOI: 10.3390/antibiotics9100712

[68] Gillis R, Adams G, Besong D, Machová E, Ebringerová A, Rowe A, et al. Application of novel analytical ultracentrifuge analysis to solutions of fungal mannans. European Biophysics Journal. 2017;**46**(**3**):235-245. DOI: 10.1007/s00249-016-1159-5

[69] Wells M, Potin P, Craigie J, Raven J, Merchant S, Helliwell K, et al. Algae as nutritional and functional food sources: Revisiting our understanding. Journal of Applied Phycology. 2017;**29**(**2**):949-982. DOI: 10.1007/s10811-016-0974-5

[70] Neetoo H, Ye M, Chen H. Bioactive alginate coatings to control Listeria monocytogenes on cold-smoked salmon slices and fillets. International Journal of Food Microbiology. 2010;**136**(**3**):326-331. DOI: 10.1016/j.ijfoodmicro.2009.10.003

[71] Marianne H, Marion D, Hanne I, Solveig L, Mindaugas M, Anneluise M, et al. Persistence of foodborne pathogens and their control in primary and secondary food production chains. Food Control. 2014;**44**:92-109. DOI: 10.1016/j.foodcont.2014.03.039

[72] Patel J, Sharma M, Millner P, Calaway T, Singh M. Inactivation of *Escherichia coli* O157:H7 attached to spinach harvester blade using bacteriophage. Foodborne Pathogens and Disease. 2011;**8**(**4**):541-546. DOI: 10.1089/fpd.2010.0734

[73] Tomat D, Mercanti D, Balagué C, Quiberoni A. Phage biocontrol of enteropathogenic and Shiga toxin-producing *Escherichia coli* during milk fermentation. Letters in Applied Microbiology. 2013;**57**(**1**):3-10. DOI: 10.1111/lam.12074

[74] Kim Y, Oh S, Kim SH. Released exopolysaccharide (r-EPS) produced from probiotic bacteria reduce biofilm formation of enterohemorrhagic *Escherichia coli* O157:H7. Biochemical and Biophysical Research Communications. 2009;**379**(**2**):324-329. DOI: 10.1016/j.bbrc.2008.12.053

[75] Medellin-Pena MJ, Wang H, Johnson R, Anand S, Griffiths MW. Probiotics affect virulence-related gene expression in *Escherichia coli* O157:H7. Applied and Environmental Microbiology. 2007;**73**(**13**):4259-4267. DOI: 10.1128/AEM.00159-07

[76] Yuan W, Yuk HG. Effects of sublethal Thymol, Carvacrol, and trans-Cinnamaldehyde adaptation on virulence properties of *Escherichia coli* O157:H7. Applied and Environmental Microbiology. 2019;**85**(**14**):e00271-e00219. DOI: 10.1128/AEM.00271-19

[77] Johnson-Henry KC, Hagen KE, Gordonpour M, Tompkins TA, Sherman PM. Surface-layer protein extracts from *lactobacillus helveticus* inhibit enterohaemorrhagic *Escherichia coli* O157:H7 adhesion to epithelial cells. Cellular Microbiology. 2007;**9**(**2**):356-367. DOI: 10.1111/j.1462-5822.2006.00791.x

[78] Ma D, Tan L, Chan K, Yap W, Pusparajah P, Chuah L, et al. Resveratrol-potential antibacterial agent against foodborne pathogens. Frontiers in Pharmacology. 2018;**9**:102. DOI: 10.3389/fphar.2018.00102

[79] Jamshidi A, Seifi HA, Kooshan M. The effect of short-time microwave exposures on *Escherichia coli* O157: H7 inoculated onto beef slices. African Journal of Microbiology Research. 2010;**4**(**22**):2371-2374

[80] Guenther S, Herzig O, Fieseler L, Klumpp J, Loessner MJ. Biocontrol of *Salmonella Typhimurium* in RTE foods with the virulent bacteriophage FO1-E2. International Journal of Food Microbiology. 2012;**154**(**1-2**):66-72. DOI: 10.1016/j.ijfoodmicro.2011.12.023

[81] Hooton SP, Atterbury RJ, Connerton IF. Application of a bacteriophage cocktail to reduce *Salmonella Typhimurium* U288

contamination on pig skin. International Journal of Food Microbiology. 2011;**151(2)**:157-163. DOI: 10.1016/j.ijfoodmicro.2011.08.015

[82] Bernet MF, Brassart D, Neeser JR, Servin AL. *Lactobacillus acidophilus* LA 1 binds to cultured human intestinal cell lines and inhibits cell attachment and cell invasion by enterovirulent bacteria. Gut. 1994;**35(4)**:483-489. DOI: 10.1136/gut.35.4.483

[83] Altenhoefer A, Oswald S, Sonnenborn U, Enders C, Schulze J, Hacker J, et al. The probiotic *Escherichia coli* strain Nissle 1917 interferes with invasion of human intestinal epithelial cells by different enteroinvasive bacterial pathogens. FEMS Immunology and Medical Microbiology. 2004;**40(3)**:223-229. DOI: 10.1016/S0928-8244(03)00368-7

[84] Moshiri J, Kaur D, Hambira C, Sandala J, Koopman J, Fuchs J, et al. Identification of a small molecule anti-biofilm agent against *Salmonella enterica*. Frontiers in Microbiology. 2018;**9**:2804. DOI: 10.3389/fmicb.2018.02804

[85] Bourgeois JS, Zhou D, Thurston T, Gilchrist J, Ko D. Methylthioadenosine suppresses *Salmonella* virulence. Infection and Immunity. 2018;**86(9)**:e00429-e00418. DOI: 10.1128/IAI.00429-18

[86] Bauza-Kaszewska J, Skowron K, Paluszak Z, Dobrzanski Z, Srutek M. Effect of microwave radiation on microorganisms in fish meals. Annals of Animal Science. 2014;**14(3)**:623-636

[87] Khashayar S, Zhang L, Abbas S, Majid K, Abolghasem H, Bao H, et al. Effective control of *Shigella* contamination in different foods using a novel six-phage cocktail. Food Science and Technology. 2021;**144**:7. DOI: 10.1016/j.lwt.2021.111137

[88] Shahin K, Zhang L, Bao H, Hedayatkhah A, Soleimani-Delfan A, Komijani M, et al. An in-vitro study on a novel six-phage cocktail against multi-drug resistant-ESBL *Shigella* in aquatic environment. Letters in Applied Microbiology. 2021;**72(3)**:231-237. DOI: 10.1111/lam.13418

[89] Leon-Velarde CG, Jun J, Skurnik M. *Yersini* Phages and food safety. Viruses. 2019;**11(12)**:1105. DOI: 10.3390/v11121105

[90] Welch TJ. Characterization of a novel *Yersinia ruckeri* serotype O1-specific bacteriophage with virulence-neutralizing activity. Journal of Fish Diseases. 2020;**43(2)**:285-293. DOI: 10.1111/jfd.13124